Interfacial Electrochemistry

Second Edition

Wolfgang Schmickler · Elizabeth Santos

Interfacial Electrochemistry

Second Edition

Prof.Dr. Wolfgang Schmickler
Universität Ulm
Institut für Theoretische Chemie
O25/34
Albert-Einstein-Allee 11
89069 Ulm
Germany
wolfgang.schmickler@uni-ulm.de

Dr. Elizabeth Santos
Universidad Nacional de Córdoba
Fac. Matemática, Astronomía y Física
Instituto de Física Enrique Gaviola
IFEG-CONICET
Avenida Haya de la Torre s/n
5000 Córdoba
Ciudad Universitaria
Argentina
esantos@uni-ulm.de

First edition published by
Oxford University Press, New York, 1996, ISBN 978-0-19-508932-5.

ISBN 978-3-642-44002-1 ISBN 978-3-642-04937-8 (eBook)
DOI 10.1007/978-3-642-04937-8
Springer Heidelberg Dordrecht London New York

Cover design: KuenkelLopka GmbH

Printed on acid-free paper

Springer is part of Springer Science+Business Media (www.springer.com)

Foreword to the second edition

About 15 years ago, my personal research interest expanded from surfaces in ultra-high vacuum to surfaces in an electrolyte. This proved to be a more difficult endeavor than expected as language and concepts used in the electrochemical literature and textbooks were rather inaccessible to a solid-state physicist. Fortunately, I became aware of the first edition of the Interfacial Electrochemistry, at the time authored solely by Wolfgang Schmickler. Ever since then, the book has served as a beacon to guide me from hostile seas of electrochemistry into friendly harbors of my own scientific background and it became my standard reference, cited in all but a few of my papers on the physics of the solid/electrolyte interface. I have frequently encouraged Wolfgang Schmickler to think about a second edition to account for the considerable development of the field since 1996, and it is very pleasing to see the project realized now. In treating electrochemistry from the perspective of a theoretical physicist with a life-long devotion to the solid/electrolyte interface, the new edition is written very much in the spirit of the first one. However, the present volume is more than just an update. Due to the congenial contributions of Elizabeth Santos, the treatise has expanded considerably into the chemistry of electrochemical reactions, into experimental methods and their analysis as well as into many fields of current interest. The volume also comprises a lucid treatise on electrochemical surface processes, a field in which I had the pleasure to collaborate with Wolfgang Schmickler for years. Although it covers a large field, the book is tutorial. Each chapter features introductory notes, which outline the qualitative aspects of the topic and place them into the perspective of general concepts. Enlightening introductory chapters in the first part of the book pave the ground for understanding, be the reader a chemist, a physicist, or a chemical engineer. The book thereby pays tribute to the interdisciplinary character of modern electrochemistry with its numerous, frequently unnoticed, applications in our daily life. Because of this tutorial value and its handbook character, the new Interfacial Electrochemistry belongs on the desk of every student in the field as well as into the hands of the professional.

Harald Ibach

Foreword to the first edition

When I started working in electrochemistry the textbooks used for University courses dealt predominantly with the properties of electrolyte solutions, with only a brief attempt at discussing the processes occurring at electrodes. Things began to change with the pioneering books of Delahay and of Frumkin which discussed kinetics in a way that a chemical engineer or a physical chemist might appreciate. Very little was said about interfacial structure, despite Butler's remarkable "Electrocapillarity", which was really premature as it appeared before the research needed to support this view had developed sufficiently. This was done in the subsequent years, to a large extent for mercury electrodes, but only from a macroscopic viewpoint using electrical measurements and predominantly thermodynamic analysis. In the last two decades the possibilities of obtaining atomic scale information and of analysing it have widened to an unprecedented extent. This has been reflected in some of the recent textbooks which have appeared, but none has embraced this modern point of view more wholeheartedly than Professor Schmickler's. Coming originally from a theoretical physics background and having already collaborated in an excellent (pre-molecular) electrochemistry textbook, he is well able to expound these developments and integrate them with the earlier studies of electrode kinetics in a way which brings out the key physical chemistry in a lucid way. His own extensive contributions to modern electrochemistry ensures that the exposition is based on a detailed knowledge of the subject. I have found the book a pleasure to read and I hope that it will not only be widely used by electrochemists, but also those physical chemists, biochemists and others who need to be convinced that electrochemistry is not a "mystery best left to the professional". I hope that this book will convince them that it is a major part of physical science.

Roger Parsons

Preface

The first edition of *Interfacial Chemistry* is now 15 years old, and has been out of print for about half that time. So much has happened in electrochemistry since then, that major changes were required. Therefore, we decided to join forces, as we have in other aspects of life, and write a thoroughly revised and updated version.

The outlook is the same as in the first edition: We treat the fundamentals of electrochemistry both from a microscopic and a macroscopic point of view, focusing on metal-solution interfaces. Understanding interfaces requires a basic knowledge of the two adjoining phases; therefore we start by reviewing briefly a few fundamental properties of solids and electrolyte solutions. The rest of the chapters follows more or less a logical order, beginning with the interface in the absence of reactions, through adsorption phenomena, and to reactions of increasing complexity. Special chapters are devoted to electrode surface processes, and to liquid–liquid interfaces. We conclude with the most important electrochemical experimental techniques, treating especially the methods suited for fast reactions in some detail. To some extent this is our response to the lamentable fashion to use nothing but cyclic voltammetry for the investigation of reactions. In contrast to the first edition, we do not cover the so-called non-traditional methods, which have been developed outside of electrochemistry. They would require a separate book for an adequate treatment.

So where has there been major progress during the last 15 years? Of course, we have learnt many details about the structure of adsorbate layers and, though to a lesser extent, about reaction steps. But most of this has been incremental, and can be considered as the normal development of a healthy branch of science. Breakthroughs have occurred, in our view, in our understanding of electrocatalysis and of electrochemical surface processes, and this is reflected in this book. Self-assembled monolayers is another branch that has grown tremendously, but again this topic is too diverse to be treated in any detail. Somewhat surprisingly, there has also been significant progress in the thermodynamics of solid electrodes, a subject that had been considered as closed since the works of Grahame and Parsons. This is a purely personal list, and certainly biased by the fact that we have been heavily involved in most of these topics. But anyone is welcome to disagree and to draw up his own list.

We want to thank all of our colleagues and students who have helped us in writing this book, and CONICET Argentina for continued support. Above all, we are grateful to Harald Ibach, who, besides writing a flattering foreword, took the trouble to read the whole book and gave us excellent advice on a number of issues. As a personal note, we thank Anahi and Nahuel for keeping our life in balance. It is customary to thank one's spouse for patient support; however, our spouses showed little patience, and were critical of every line we wrote.

Finally we want to recommend a few books as supplementary reading: The electrochemical textbook that we like best is Sato's [1], but Hamann, Hamnett, Vielstich [2] is also a good, general textbook and covers applied topics as well. Ibach's monograph [3] covers the physics of surfaces and interfaces with precision, and complements ours. Of the older books, Delahay's [4] is the best, and an invaluable source for transient techniques.

May 2010 Elizabeth Santos and Wolfgang Schmickler

References

1. N. Sato, *Electrochemistry at Metal and Semiconductor Electrodes*, Elsevier, Amsterdam, 1998.
2. C. H. Hamann, A. Hamnett, and W. Vielstich, *Electrochemistry*, Wiley-VCH, Weinheim, 2007.
3. H. Ibach, *Physics of Surfaces and Interfaces*, Springer, Berlin, Heidelberg, 2006.
4. P. Delahay, *Double Layer and Electrode Kinetics*, Interscience, New York, NY, 1966.

Contents

1

Introduction

1.1 The scope of electrochemistry

Electrochemistry is an old science: There is good archaeological evidence that an electrolytic cell was used by the Parthans (250 B.C. to 250 A.D.), probably for electroplating (see Fig. 1.1), though a proper scientific investigation of electrochemical phenomena did not start before the experiments of Volta and Galvani [1, 2]. The meaning and scope of electrochemical science has varied throughout the ages: For a long time it was little more than a special branch of thermodynamics; later attention turned to electrochemical kinetics. During recent decades, with the application of various surface-sensitive techniques to electrochemical systems, it has become a science of interfaces, and this, we think, is where its future lies. There are a large variety of interfaces of interest to electrochemists, and Fig. 1.2 shows several examples. So in this book we use as a working definition:

Fig. 1.1. Remnants of a cell used by the Parthans: It consists of an iron core surrounded by a copper cylinder, both immersed in a clay jar.

W. Schmickler, E. Santos, *Interfacial Electrochemistry*, 2nd ed.,
DOI 10.1007/978-3-642-04937-8_1,

Electrochemistry is the study of structures and processes at the interface between an electronic conductor (the electrode) *and an ionic conductor* (the electrolyte) *or at the interface between two electrolytes.*

This definition requires some explanation. (1) By *interface* we denote those regions of the two adjoining phases whose properties differ significantly from those of the bulk. These interfacial regions can be quite extended, particularly in those cases where a metal or semiconducting electrode is covered by a thin film or an adsorbate layers. The modification of the electrode surface by different types of adsorbates (Fig. 1.2 bottom) can produce very complicated structures. Such modified electrodes have important applications in different fields, such as protection against corrosion, in electrocatalysis, and the development of sensors. Sometimes the term *interphase* is used to indicate the spatial extention. (2) It would have been more natural to restrict the definition to the interface between an electronic and an ionic conductor only, and, indeed, this is generally what we mean by the term *electrochemical interface*. However, the study of the interface between two immiscible electrolyte solutions is so similar that it is natural to include it under the scope of electrochemistry.

Metals and semiconductors are common examples of electronic conductors, and under certain circumstances even insulators can be made electronically conducting, for example by photoexcitation. Electrolyte solutions, molten salts including ionic liquids, and solid electrolytes are ionic conductors. Some

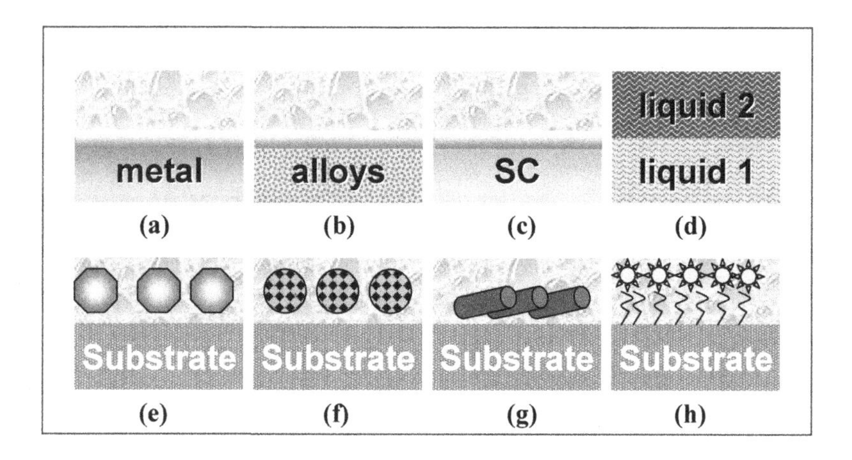

Fig. 1.2. A few important types of electrochemical interfaces. *Top*: (**a**) metal/electrolyte; (**b**) alloy/electrolyte; (**c**) semiconductor/electrolyte; (**d**) two immiscible liquids in contact. *Bottom*: The electrode (substrate) has been modified by deposition of different adsorbates: (**e**) nanoparticles; (**f**) fullerenes; (**g**) nanotubes; (**h**) functionalized self assembled monalayer.

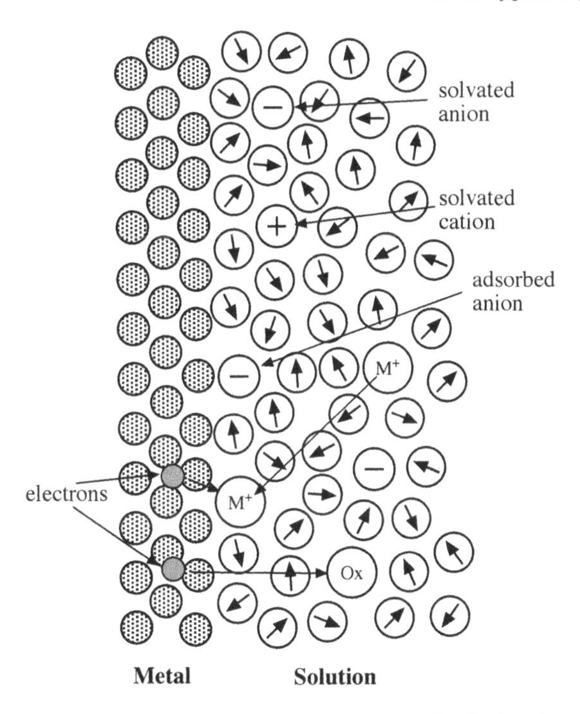

Fig. 1.3. Structure and processes at the metal-solution interface.

materials have appreciable electronic *and* ionic conductivities, and depending on the circumstances one or the other or both may be important.

With metals, semiconductors, and insulators as possible electrode materials, and solutions, molten salts, and solid electrolytes as ionic conductors, there is a fair number of different classes of electrochemical interfaces. However, not all of these are equally important: The majority of contemporary electrochemical investigations is carried out at metal-solution or at semiconductor-solution interfaces. We shall focus on these two cases, and consider some of the others briefly.

1.2 A typical system: the metal-solution interface

To gain an impression of the structures and reactions that occur in electrochemical systems, we consider the interface between a metal and an electrolyte solution. Figure 1.3 shows a schematic diagram of its structure. Nowadays most structural investigations are carried out on single crystal surfaces; so the metal atoms, indicated by the dotted circles on the left, are arranged in a lattice. Solvent molecules generally carry a dipole moment, and are hence represented as spheres with a dipole moment at the center. Ions are indicated by spheres with a charge at the center. Near the top of the picture we observe an anion and a cation, which are close to the electrode surface but not in contact with it. They are separated from the metal by their solvation sheaths. A little below is an anion in contact with the metal; we say it is *specifically adsorbed* if it is held there by chemical interactions. Usually anions are less strongly solvated than cations; therefore their solvation sheaths are easier to break up, and they are more often specifically adsorbed, particularly on positively charged metal surfaces. Adsorption occurs on specific sites; the depicted anion is adsorbed on top of a metal atom, in the *atop* position. The two types of reactions shown near the bottom of the figure will be discussed below.

Generally the interface is charged: the metal surface carries an excess charge, which is balanced by a charge of equal magnitude and opposite sign on the solution side of the interface. Figure 1.4 shows the charge distribution for the case in which the metal carries a positive excess charge, and the solution a negative one – there is a deficit of electrons on the metal surface, and more anions than cations on the solution side of the interface. Since a metal electrode is an excellent conductor, its excess charge is restricted to a surface region about 1 Å thick. Usually one works with fairly concentrated (0.1–1 M) solutions of strong electrolytes. Such solutions also conduct electric currents well, though their conductivities are several orders of magnitude smaller than those of metals. For example, at room temperature the conductivity of silver is $0.66 \times 10^6\ \Omega^{-1}$cm; that of a 1 M aqueous solution of KCl is $0.11\ \Omega^{-1}$cm. The greater conductivity of metals is caused both by a greater concentration of charge carriers and by their higher mobilities. Thus silver has an electron concentration of 5.86×10^{22} cm^{-3}, while a 1 M solution of KCl has about 1.2×10^{21} ions cm^{-3}. The difference in the mobilities of the charge carriers is thus much greater than the difference in their concentrations. Because of the lower carrier concentration, the charge in the solution extends over a larger region of space, typically 5–20 Å thick. The resulting charge distribution – two narrow regions of equal and opposite charge – is known as *the electric double layer*. It can be viewed as a capacitor with an extremely small effective plate separation, and therefore has a very high capacitance.

The voltage drop between the metal and the solution is typically of the order of 1 V. If the voltage is substantially higher, the solution is decomposed – in aqueous solutions either oxygen or hydrogen evolution sets in. Since this potential drop extends over such a narrow region, it creates extremely high fields of up to 10^9 Vm^{-1}. Such a high field is one of the characteristics of electrochemical interfaces. In vacuum fields of this magnitude can only be generated at sharp tips and are therefore strongly inhomogeneous. Electrochemical experiments on metals and semiconductors are usually performed with a time resolution of 1 μs or longer[1] – a few milliseconds is typical for

[1] For the following reason: electrochemical experiments involve a change of the electrode potential, and hence charging or discharging the capacitor formed by the double layer. Since the double-layer capacity is large, and the resistance of

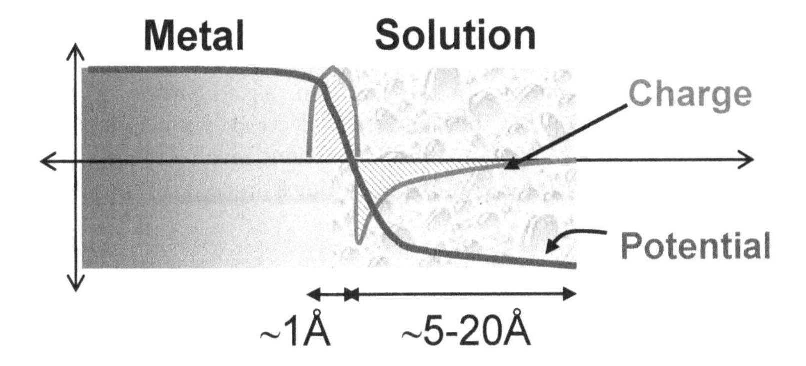

Fig. 1.4. Distribution of charge and potential at the metal-solution interface (schematic).

transient measurements (details will be given in Chap. 13). If one looks at the interface over this time range, the positions of the ions are smeared out, and one only sees a homogeneous charge distribution and hence a homogeneous electrostatic field. Inhomogeneities may exist near steps, kinks, or similar features on the metal surface.

The structure of the interface is of obvious interest to electrochemists. However, the interface forms only a small part of the two adjoining phases, and spectroscopic methods which generate signals both from the bulk and from the interface are not suitable for studying the interface, unless one finds a way of separating the usually dominant bulk signal from the small contribution of the interface. Techniques employing electron beams, which have provided a wealth of data for surfaces in the vacuum, cannot be used since electrons are absorbed by solutions. Indeed, a lack of spectroscopic methods that are sensitive to the interfacial structure has for a long time delayed the development of electrochemistry, and only the past 20–30 years have brought substantial progress.

Reactions involving charge transfer through the interface, and hence the flow of a current, are called *electrochemical reactions.* Two types of such reactions are indicated in Fig. 1.3. The upper one is an instance of *metal deposition.* It involves the transfer of a metal ion from the solution onto the metal surface, where it is discharged by taking up electrons. Metal deposition takes place at specific sites; in the case shown it is a *hollow site* between the atoms of the metal electrode. The deposited metal ion may belong to the same species as those on the metal electrode, as in the deposition of a Ag^+ ion on a silver electrode, or it can be different as in the deposition of a Ag^+ ion on platinum. In any case the reaction is formally written as:

---

the solution is not negligible, it has a long time constant associated with it, and the response at short times is dominated by this charging of the double layer.

$$\mathrm{Ag^+(solution) + e^-(metal) \rightleftharpoons Ag(metal)} \tag{1.1}$$

Metal deposition is an example of a more general class of electrochemical reactions, *ion-transfer reactions*. In these an ion, e.g. a proton or a chloride ion, is transferred from the solution to the electrode surface, where it is subsequently discharged. Many ion-transfer reactions involve two steps. The hydrogen-evolution reaction, for example, sometimes proceeds in the following way:

$$\mathrm{H_3O^+ + e^- \rightleftharpoons H_{ad} + H_2O} \tag{1.2}$$

$$\mathrm{2H_{ad} \rightleftharpoons H_2} \tag{1.3}$$

where H_{ad} refers to an adsorbed proton. Only the first step is an electrochemical reaction; the second step is a purely chemical recombination and desorption reaction.

Another type of electrochemical reaction, an *electron-transfer reaction*, is indicated near the bottom of Fig. 1.3. In the example shown an oxidized species is reduced by taking up an electron from the metal. Since electrons are very light particles, they can tunnel over a distance of 10 Å or more, and the reacting species need not be in contact with the metal surface. The oxidized and the reduced forms of the reactants can be either ions or uncharged species. A typical example for an electron-transfer reaction is:

$$\mathrm{Fe^{3+}(solution) + e^-(metal) \rightleftharpoons Fe^{2+}(solution)} \tag{1.4}$$

Both ion- and electron-transfer reactions entail the transfer of charge through the interface, which can be measured as the electric current. If only one charge transfer reaction takes place in the system, its rate is directly proportional to the current density, i.e. the current per unit area. This makes it possible to measure the rates of electrochemical reactions with greater ease and precision than the rates of chemical reactions occurring in the bulk of a phase. On the other hand, electrochemical reactions are usually quite sensitive to the state of the electrode surface. Impurities have an unfortunate tendency to aggregate at the interface. Therefore electrochemical studies require extremely pure system components.

Since in the course of an electrochemical reaction electrons or ions are transferred over some distance, the difference in the electrostatic potential enters into the Gibbs energy of the reaction. Consider the reaction of Eq. (1.4), for example. For simplicity we assume that the potential in the solution, at the position of the reacting ion, is kept constant. When the electrode potential is changed by an amount $\Delta\phi$, the Gibbs energy of the electron is lowered by an amount $-e_0\,\Delta\phi$, and hence the Gibbs energy of the reaction is raised by $\Delta G = e_0\,\Delta\phi$. Varying the electrode potential offers a convenient way of controlling the reaction rate, or even reversing the direction of a reaction, again an advantage unique to electrochemistry.

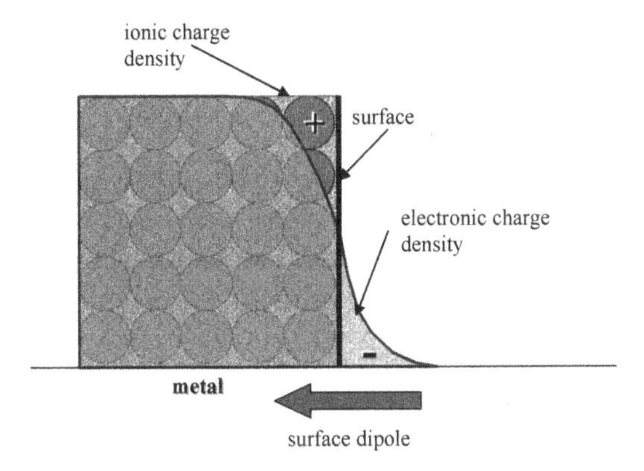

Fig. 1.5. Charge distribution and surface dipole at a metal surface. For simplicity the positive charge residing on the metal ions has been smeared out into a constant background charge.

1.3 Inner, outer, and surface potentials

Electrochemical interfaces are sometimes referred to as *electrified interfaces*, meaning that potential differences, charge densities, dipole moments, and electric currents occur. It is obviously important to have a precise definition of the electrostatic potential of a phase. There are two different concepts. The *outer* or *Volta potential* ψ_α of the phase α is the work required to bring a unit point charge from infinity to a point just outside the surface of the phase. By "just outside" we mean a position very close to the surface, but so far away that the image interaction with the phase can be ignored; in practice, that means a distance of about $10^{-5} - 10^{-3}$ cm from the surface. Obviously, the outer potential ψ_α is a measurable quantity.

In contrast, the *inner* or *Galvani potential* ϕ_α is defined as the work required to bring a unit point charge from infinity to a point *inside* the phase α; so this is the electrostatic potential which is actually experienced by a charged particle inside the phase. Unfortunately, the inner potential cannot be measured: If one brings a real charged particle – as opposed to a point charge – into the phase, additional work is required due to the chemical interaction of this particle with other particles in the phase. For example, if one brings an electron into a metal, one has to do not only electrostatic work, but also work against the exchange and correlation energies.

The inner and outer potential differ by the *surface potential* $\chi_\alpha = \phi_\alpha - \psi_\alpha$. This is caused by an inhomogeneous charge distribution at the surface. At a metal surface the positive charge resides on the ions which sit at particular lattice sites, while the electronic density decays over a distance of about 1 Å from its bulk value to zero (see Fig. 1.5). The resulting dipole potential is of the order of a few volts and is thus by no means negligible. Smaller surface potentials exist at the surfaces of polar liquids such as water, whose molecules have a dipole moment. Intermolecular interactions often lead to a small net orientation of the dipoles at the liquid surface, which gives rise to a corresponding dipole potential.

The inner potential ϕ_α is a bulk property. Even though it cannot be measured, it is still a useful concept, particularly for model calculations. Differences in the inner potential of two phases can be measured, if they have the same chemical composition. The surface potential χ_α is a surface property, and may hence differ at different surfaces of a single crystal. The same is then also true of the outer potential ψ; thus different surface planes of a single crystal of a metal generally have different outer potentials. We will return to these topics below.

Problems

1. Consider the surface of a silver electrode with a square arrangement of atoms (this is a so-called Ag(100) surface, as will be explained in Chap. 4) and a lattice constant of 2.9 Å. (a) What is the excess-charge density if each Ag atom carries an excess electron? (b) How large is the resulting electrostatic field if the solution consists of pure water with a dielectric constant of 80? (c) In real systems the excess-charge densities are of the order of ± 0.1 C m^{-2}. What is the corresponding number of excess or defect electrons per surface atom? (d) If a current density of 0.1 A cm^{-2} flows through the interface, how many electrons are exchanged per second and per silver atom?
2. Consider a plane metal electrode situated at $z = 0$, with the metal occupying the half-space $z \leq 0$, the solution the region $z > 0$. In a simple model the excess surface charge density σ in the metal is balanced by a space charge density $\rho(z)$ in the solution, which takes the form: $\rho(z) = A\exp(-\kappa z)$, where κ depends on the properties of the solution. Determine the constant A from the charge balance condition. Calculate the interfacial capacity assuming that κ is independent of σ.
3. In a simple model a water molecule is represented as a hard sphere with a diameter $d = 3$ Å and a dipole moment $m = 6.24 \times 10^{-30}$ Cm at its center. Calculate the energy of interaction E_{int} of a water molecule with an ion of radius a for the most favorable configuration. When an ion is adsorbed, it loses at least one water molecule from its solvation shell. If the ion keeps its charge it gains the image energy E_{im}. Compare the magnitudes of E_{int} and E_{im} for $a = 1$ and 2 Å. Ignore the presence of the water when calculating the image interaction.

References

1. A. Volta, *Phil. Trans.* **II** (1800) 405–431; *Gilbert's Ann.* **112** (1800) 497.
2. A. Galvani, De Viribus Electricitatis in Motu Musculari Commentarius, *ex Typ. Instituti Scientiarum Bononiae*, 1791; see also: S. Trasatti, *J. Electroanal. Chem.* **197** (1986) 1.

2

Metal and semiconductor electrodes

Before treating processes at the electrochemical interface, it is useful to review a few basic properties of the adjoining phases, the electrolyte and the electrode. So here we summarize important properties of metals and semiconductors. Liquid electrolyte solutions, which are the only electrolytes we consider in this book, will be treated in the next chapter. These two chapters are not meant to serve as thorough introductions into the physical chemistry of condensed phases, but present the minimum that a well-educated electrochemist should known about solids and solutions.

2.1 Metals

In a solid, the electronic levels are not discrete like in an atom or molecule, but they form bands of allowed energies. In an elemental solid, these bands are formed by the overlap of like orbitals in neighboring atoms, and can therefore be labeled by the orbitals of which they are composed. Thus, we can speak of a $1s$ or a $3d$ band. The bands are the wider, the greater the overlap between the orbitals. Therefore the bands formed by the inner electron levels are narrow; they have low energies and generally play no role in bonding or in chemical reactions. The important bands are formed by the valence orbitals, and they are of two types: the s and p orbitals tend to have similar energies, they overlap well, and they form broad sp bands. In contrast, the d orbitals are more localized, their overlap is smaller, and they form rather narrow d bands.

At $T = 0$ the bands are filled up to a certain level, the *Fermi level* E_F. It is a characteristic of metals that the Fermi level lies inside an energy band, which is therefore only partially filled. This is the reason why metals are good conductors, because neither empty not completely filled bands contribute to the conductivity.[1] At finite temperatures, electrons can be excited thermally

[1] The latter fact may seem a little surprizing. The actual proof is not simple, but, naively speaking, the electrons cannot move because they have nowhere to go.

W. Schmickler, E. Santos, *Interfacial Electrochemistry*, 2nd ed.,
DOI 10.1007/978-3-642-04937-8_2, © Springer-Verlag Berlin Heidelberg 2010

to levels above the Fermi level, leaving behind an unoccupied state or *hole*. The distribution of electrons and holes is restricted to an energy region of a few $k_B T$ around the E_F. Quantitatively, the probability that an energy level of energy ϵ is filled, is given by the Fermi–Dirac distribution depicted in Fig. 2.1.:

$$f(\epsilon) = \frac{1}{1 + \exp(\frac{\epsilon - E_F)}{k_B T}} \tag{2.1}$$

Strictly speaking, this equation should contain the electrochemical potential of the electrons instead of the Fermi level, but for metal near room temperature, which we consider here, the difference is negligible.

At room temperature, $k_B T \approx 0.025$ eV; often energies of this order of magnitude are negligible, and the Fermi–Dirac distribution can then be replaced by a step function:

$$f(\epsilon) \approx H(E_F - \epsilon), \qquad H(x) = \begin{cases} 1 \text{ for } x > 0 \\ 0 \text{ for } x \leqslant 0 \end{cases} \tag{2.2}$$

For high energies the Fermi–Dirac distribution goes over into the Boltzmann distribution:

$$f(\epsilon) \approx \exp -\frac{\epsilon - E_F}{k_B T} \qquad \text{for} \qquad \epsilon \gg E_F \tag{2.3}$$

We also note the following symmetry between the probability of finding an occupied and an empty state (hole):

$$1 - f(\epsilon) = f(-\epsilon) \tag{2.4}$$

The distribution of the electronic levels within a band is given by the *density of states* (DOS). In electrochemistry, the DOS at the surface is of primary importance. It differs somewhat from the DOS in the bulk because of the different coordination of the surface atoms. Figure 2.2 shows the DOS at

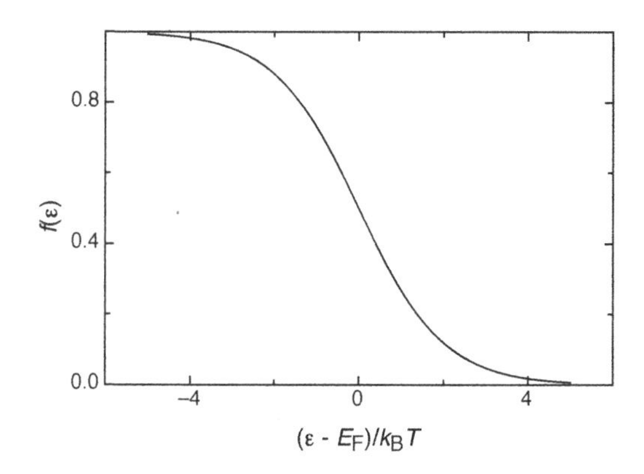

Fig. 2.1. Fermi–Dirac distribution.

the (111) surface of silver – the notation will be explained in the next section. The *sp* band is wide and has a pronounced maximum near −6 eV below the Fermi level, which is mostly due to the *s* states. In contrast, the *d* band is narrow and ends several eV below the Fermi level. We will see later, that this distribution of the *d* band has a significant effect on the catalytic properties of silver.

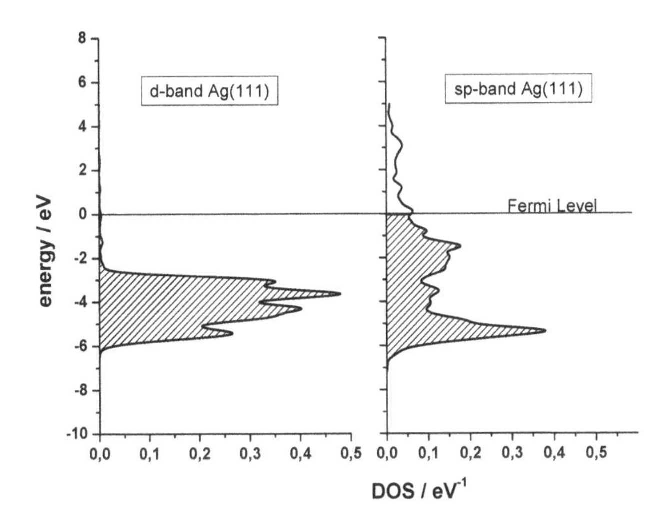

Fig. 2.2. Densities of state for the *d* band and the *sp* band at the Ag(111) surface. Their integrals has been normalized to unity, and the Fermi level has been taken as the energy zero.

2.2 Single crystal surfaces

The structure of electrode surfaces is of primary importance for electrochemistry. Fundamental research is nowadays mostly done on single crystals, which have a simple and well-defined surface structure. Many metals that are used in electrochemistry (Au, Ag, Cu, Pt, Pd, Ir) have a *face-centered cubic* (fcc) lattice, so we will consider this case in some detail. For other lattice structures we refer to the pertinent literature and to Problem 1.

Figure 2.3 shows a conventional unit cell of an fcc crystal. It consists of atoms at the eight edges of a cube and at the centers of the six sides. The length of the side of the cube is the *lattice constant*; for our present purpose we may assume that it is unity. The lattice of an infinite, perfect solid is obtained by repeating this cubic cell periodically in all three directions of space.

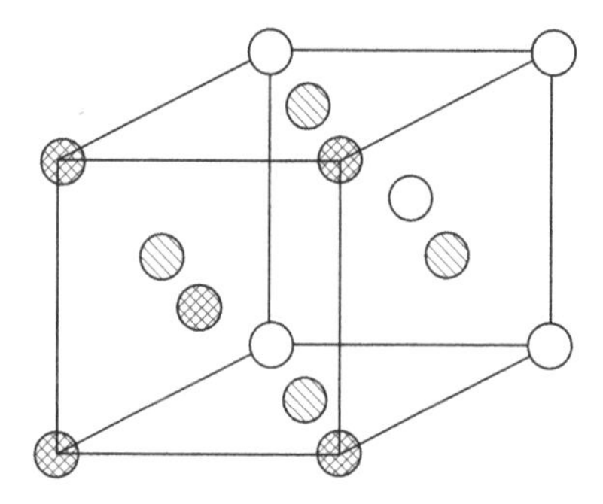

Fig. 2.3. Conventional unit cell of a face-centered cubic crystal. The lattice contains the points at the corners of the *cube* and the points at the centers of the six *sides*.

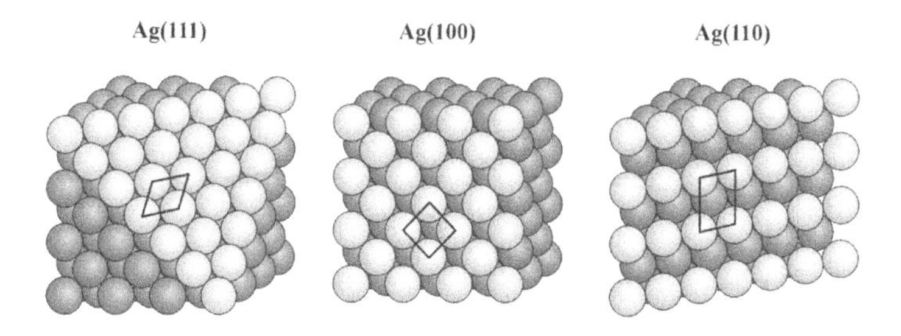

Fig. 2.4. The principal lattice planes of a face-centered cubic crystal and the principal lattice planes.

A perfect surface is obtained by cutting the infinite lattice in a plane that contains certain lattice points, a *lattice plane* (Fig. 2.4). The resulting surface forms a two-dimensional sublattice, and we want to classify the possible surface structures. Parallel lattice planes are equivalent in the sense that they contain identical two-dimensional sublattices, and give the same surface structure. Hence we need only specify the direction of the normal to the surface plane. Since the length of this normal is not important, one commonly specifies a normal vector with simple, integral components, and this uniquely specifies the surface structure.

For an fcc lattice a particularly simple surface structure is obtained by cutting the lattice parallel to the sides of a cube that forms a unit cell (see Fig. 2.5a) . The resulting surface plane is perpendicular to the vector (1,0,0); so this is called a (100) surface, and one speaks of Ag(100), Au(100), etc., surfaces, and (100) is called the *Miller index*. Obviously, (100), (010), (001) surfaces have the same structure, a simple square lattice, whose lattice constant is $a/\sqrt{2}$. Adsorption of particles often takes place at particular surface sites, and some of them are indicated in the figure: The position on top of a

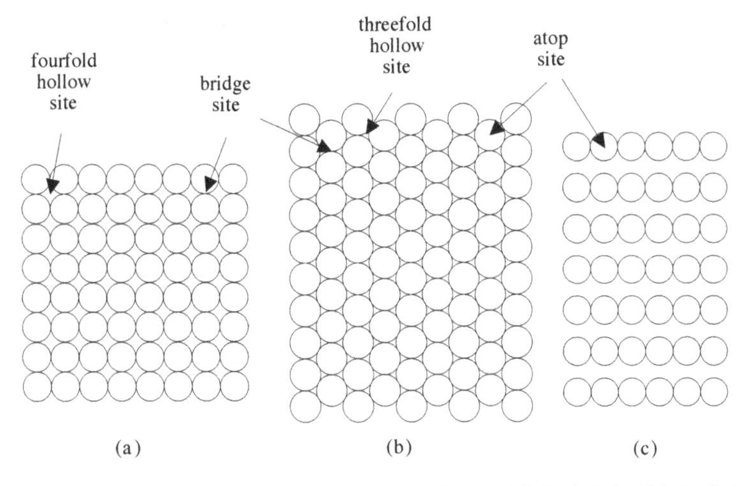

Fig. 2.5. Lattice structures of single crystal surfaces: (**a**) fcc(100), (**b**) fcc(111), (**c**) fcc(110).

lattice site is the *atop* position, *fourfold hollow sites* are in the center between the surface atoms, and *bridge sites* (or *twofold hollow sites*) are in the center of a line joining two neighboring surface atoms.

The densest surface structure is obtained by cutting the lattice perpendicular to the [111] direction (see Fig. 2.5b). The resulting (111) surface forms a triangular (or hexagonal) lattice and the lattice constant is $a/\sqrt{2}$. Important sites for adsorption are the *atop*, the *bridge*, and the *threefold hollow* sites (Fig. 2.5b).

The (110) surface has a lower density than either the (111) or the (100) planes (Fig. 2.5c). It forms a rectangular lattice; the two sides of the rectangle are a and $a/\sqrt{2}$. The resulting structure has characteristic grooves in one direction.

The three basal planes, (100), (111) and (110), define the vertices of a stereographic triangle [2]. When these surfaces are appropriately treated by annealing, they show large, highly uniform terraces. However, the catalytic activity is sometimes better at defect sites. The simplest example of a defect is a vacancy or its opposite, an adatom. In addition, dislocations in the bulk propagating outside of the crystal produce mesoscopic defects, which appear as steps at the surface.

An interesting method to systematically investigate mesoscopic defects is to cut the crystal at a small angle θ with respect to one of the basal planes to expose a high index plane consisting of terraces of low index planes, with constant width, linked by steps often of monoatomic height. The terraces can extend to large distances in a given direction of the crystal. These surfaces are called *vicinal surfaces* and the terrace/step geometry is determined by the cutting angle. In Fig. 2.6 we show three different stepped surfaces, the (997),

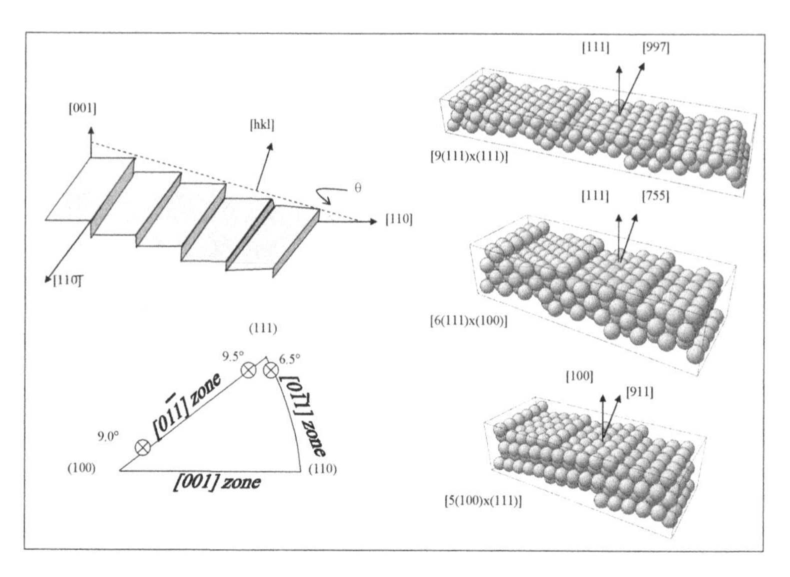

Fig. 2.6. Examples of vicinal surfaces. A *bar* across a Miller index indicates a negative number.

the (755) and the (911). They were obtained cutting the crystal at 6.5° and 9.5° with respect to the (111) plane for the two first, but towards different directions, and at 9.0° with respect to the (100) plane for the latter. The (997) and the (755) have (111) terraces of different lengths, and monoatomic (111) and (100) steps, respectively. A more convenient nomenclature for these high index faces, which indicates better their structures, is that proposed by Lang et al. [1] and it is also given in the figure. This is equivalent to a high Miller index and has the form: $[m(hkl) \times n(h'k'l')]$, where the first part designates a terrace of Miller index (hkl) with m infinite atomic rows and the second part indicates a step of Miller index $(h'k'l')$ and n atomic layers high. Obviously, this are nominal structures; depending on their thermal stability, they may undergo reconstruction (see Chap. 16).

2.3 Semiconductors

Electronic states in a perfect semiconductor are delocalized just as in metals, and there are bands of allowed electronic energies. In semiconductors the current-carrying bands do not overlap as they do in metals; they are separated by the *band gap*, and the Fermi level lies right in this gap (see Fig. 2.7).

The band below the Fermi level, which at $T = 0$ is completely filled, is known as the *valence band*; the band above, which is empty at $T = 0$, is the *conduction band*. In a pure or *intrinsic semiconductor*, the Fermi level is close to the center of the band gap. At room temperature a few electrons

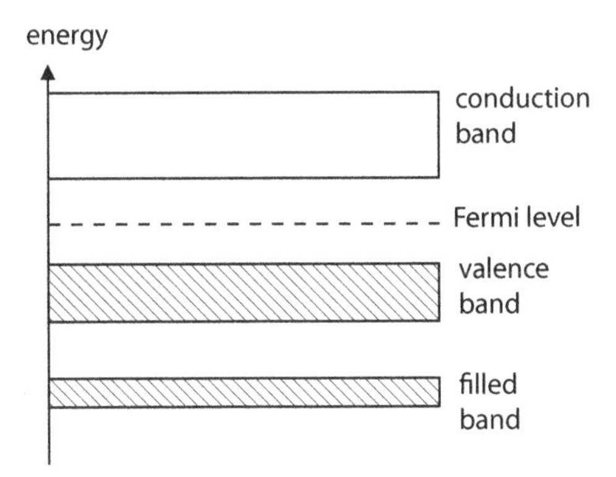

Fig. 2.7. Band structure of an intrinsic semiconductor. At $T = 0$ the valence band is completely filled and the conduction band is empty. At higher temperatures the conduction band contains a low concentration of electrons, the valence band an equal concentration of holes. Bands with a lower energy, one of which is shown, are always completely filled.

are excited from the valence into the conduction band, leaving behind electron vacancies or *holes* (denoted by h^+). The electric current is carried by electrons in the conduction band and holes in the valence band. Just like in metals, the concentrations n_c of the conduction electrons and p_v of the holes are also governed Fermi statistics. Denoting by E_c the lower edge of the conduction band, and by N_c the effective density of states at E_c, the concentration of electrons is:

$$n_c = N_c f(E_c - E_F) \approx N_c \exp\left(-\frac{E_c - E_F}{kT}\right) \tag{2.5}$$

The last approximation is valid if $E_c - E_F \gg kT$ (i.e., if the band edge is at least a few kT above the Fermi level), and the Fermi–Dirac distribution $f(\epsilon)$ can be replaced by the Boltzmann distribution. Similarly, the concentration of holes in the valence band is:

$$p_v = N_v \left[1 - f(E_v - E_F)\right] \approx N_v \exp\left(-\frac{E_F - E_v}{kT}\right) \tag{2.6}$$

where E_v is the upper edge of the valence band, and N_v the effective density of states at E_v. The last approximation is valid if $E_F - E_v \gg kT$. If the Fermi level lies within a band, or is close (i.e. within kT) to a band edge, one speaks of a *degenerate semiconductor*.

The band gap E_g of semiconductors is typically of the order of 0.5–2 eV (e.g., 1.12 eV for Si, and 0.67 eV for Ge at room temperature), and consequently the conductivity of intrinsic semiconductors is low. It can be greatly enhanced by *doping*, which is the controlled introduction of suitable impurities. There are two types of dopants: *Donors* have localized electronic states with energies immediately below the conduction band, and can donate their

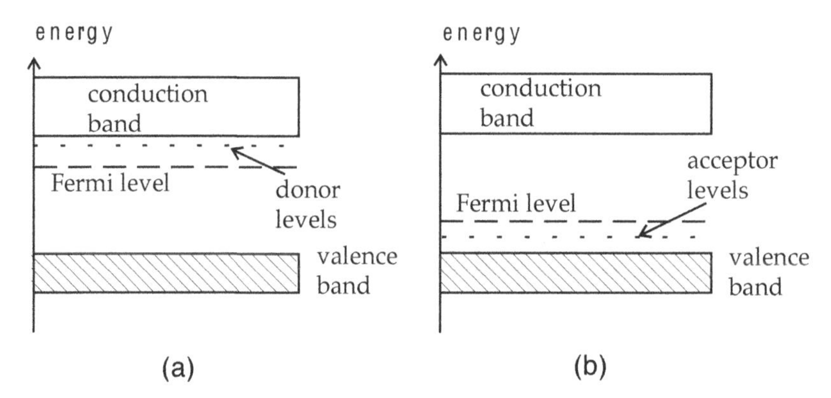

Fig. 2.8. Band structure of (**a**) an n-type and (**b**) a p-type semiconductor with fully ionized donors.

electrons to the conduction band; in accord with Eq. (2.5) this raises the Fermi level toward the lower edge of the conduction band (see Fig. 2.8a). Semiconductors with an excess of donors are n-type , and the electrons constitute the *majority carriers* in this case, and the holes are the *minority carriers.* In contrast, *acceptors* have empty states just above the valence band, which can accept an electron from the valence band, and thus induce holes. Consequently, the Fermi level is shifted toward the valence band (see Fig. 2.8b); we speak of a p-type semiconductor, and the holes constitute the majority, the electrons the minority carriers.

2.4 Comparison of band structures

Figure 2.9 shows schematically the band structure of a few typical electrode materials, three metals (platinum, gold and silver) and a semiconductor (silicon). All three metals possess a wide *sp* band extending well above the Fermi level. However, the d bands are different. The position of the d band of silver is lower than that of gold, and both lie lower than that of platinum. In the latter case the d band even extend about 0.5 eV above the Fermi level. As we shall see later, these differences are crucial for the electrocatalytical properties of these materials.

According to the Fermi distribution, all electronic states below the Fermi level are occupied for the four materials, although for the metals a small numbers of electrons can be excited thermally within an energy range of about kT around the Fermi level. This effect is represented by the shadowing near the E_F. In the case of the semiconductor, the band gap of undoped silicon is too large to allow the electrons to be excited thermally into the conduction band, since the band gap E_g is much wider than the thermal energy.

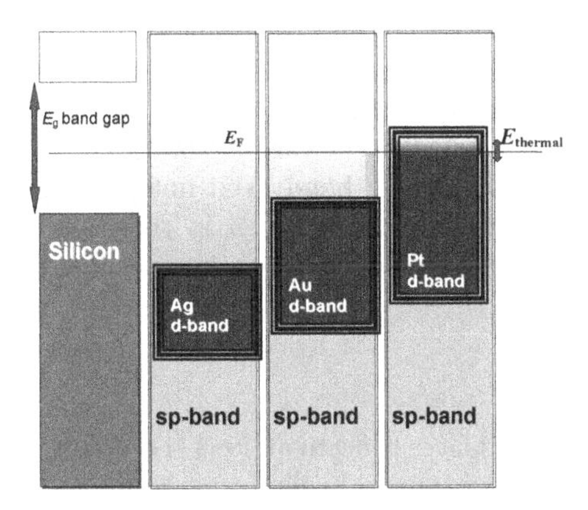

Fig. 2.9. Band structures of a semiconductor and a few metals (schematic). The energies are not to scale.

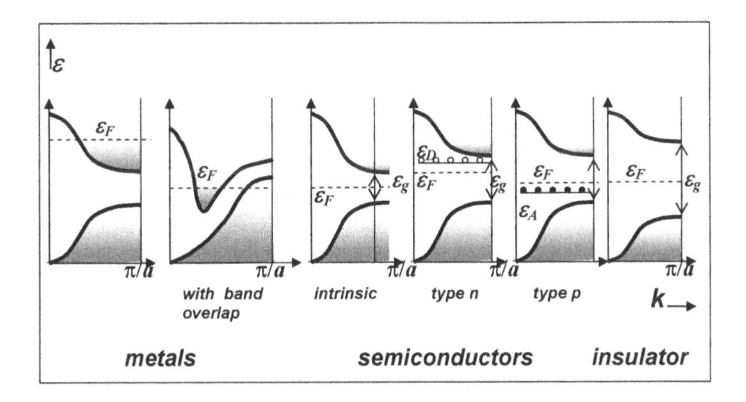

Fig. 2.10. One-dimensional model for few typical band structures as a function of the wave-vector k.

Figure 2.9 just shows the allowed energy levels, but contains no information about the wave-functions. In a solid, the electronic wavefunctions depend on the wavenumber $\mathbf{k}$. In the simple free electron model, the corresponding wavefunctions are simply plane waves of the form $\exp \mathbf{kx}$, with momenta $\hbar\mathbf{k}$ and energy $E_{\mathbf{k}} = \hbar^2 k^2/2m$. However, in a real crystal the electrons experience the three-dimensional periodic potential of the nuclei. While the vector $\mathbf{k}$ can still be used as a quantum number, the expression for the energy is no longer simple. We shall need this $\mathbf{k}$ dependence of the energy only in Chap. 11, when we consider optical excitations. For a basic understanding it is sufficient to consider a one-dimensional case, in which the electrons experience a periodic potential with lattice constant a. The periodicity in space induces a corresponding periodicity in k, and it is sufficient to consider values of k in the range $[0, \pi/a]$, and plot $E(k)$ in this range.

Figure 2.10 shows a few typical cases. Note that the minimum of the conduction band and the valence band need not coincide, as in the second figure from the left. This can also happen in semiconductors, and will be treated in Chap. 11.

Problems

1. (a) Consider the second layer beneath an fcc(111) surface and verify, that there are two different kinds of threefold hollow sites on the surface. (b) The conventional unit cell of a body-centered cubic (bcc) lattice consists of the corners and the center of a cube. Determine the structures of the bcc(111), bcc(100), and bcc(110) surfaces.
2. *One-dimensional free electron gas* We consider a simple model for a one-dimensional solid. It is represented by a box extending between $x = 0$ and $x = L$ with infinite walls. This is a well-known problem in quantum mechanics. Show that the wavefunctions have the form:

$$\phi_n(x) = \left(\frac{2}{L}\right)^{1/2} \sin(n\pi x/L), \quad n \in \mathbb{N} \tag{2.7}$$

with an energy:

$$\epsilon_n = \frac{h^2}{2m}\left(\frac{n}{2L}\right)^2 \tag{2.8}$$

Let N be the total number of electrons in the solid, which we can take to be an even number. Using the fact, that each level n can be occupied by two electrons of opposite spin, show that the Fermi energy is:

$$E_{\mathrm{F}} = \frac{h^2}{2m}\left(\frac{N}{4L}\right)^2 \tag{2.9}$$

Calculate the Fermi energy for the case where the density per length is $N/L = 0.5$ electrons per Å. Show that for sufficiently large N the total energy of the ground state (at $T = 0$) is given by:

$$E_0 = \frac{1}{3} N E_{\mathrm{F}} \tag{2.10}$$

In the limit of $L \to \infty$, the quantum number n becomes continuous. The density of states ρ_ϵ is the number of electron states per unit of energy. Using Eq. (2.8), show that:

$$\rho(\epsilon) = 2\frac{dn}{d\epsilon} = \frac{4L}{h}\left(\frac{m}{2\epsilon}\right)^{1/2} \tag{2.11}$$

References

1. B. Lang, R.W. Joyner, and G.A. Somorjai, *Surf. Sci.* **30** (1972) 440.
2. A. Hamelin and J. Lecoeur, *Surf. Sci.* **57** (1976) 771.

3

Electrolyte solutions

3.1 The structure of water

Most solvents are polar, i.e. their molecules have a permanent dipole moment; a few typical values are listed in Table 3.1. For comparison we note that the dipole moment of two unit charges, of opposite sign and at a separation of 1 Å, has a value of 1.6×10^{-29} Cm. Good solvents typically have dipole moments that are only a little lower. The resulting strong electrostatic interaction between the molecules is usually the reason, why these substances are liquid at ambient temperatures. However, there are also good solvents, such as CCl_4, which have no or only a small permanent dipole moment but a high polarizability, so that the presence of ionic charges induces a sizable dipole moment.

Water is by far the most important solvent [1], and is used in most electrochemical experiments, although it does have one disadvantage: The potential window between hydrogen and oxygen evolution is only about 1.2 V, and is thus smaller than that of several non-aqueous solvents, which have stability ranges of up to 4 V. However, non-aqueous solvents are difficult to handle, are not so healthy as water, and not popular with experimentalist.

Water molecules not only interact which each other through their dipole moments, but in addition form hydrogen bonds, which increases their cohesion. The molecules of liquid water form a network of molecules, which is held

Solvent	$10^{-30} \cdot \mu/\mathrm{Cm}$	$10^{-30} \cdot \alpha/\ \mathrm{m}^3$
H_2O	6.14	1.5
HCl	3.44	2.6
NH_3	4.97	2.26
CCl_4	4.97	2.26

Table 3.1. Dipole moment μ and polarizability α of a few solvents; note: 10^{-30} Cm= 0.3 Debye.

W. Schmickler, E. Santos, *Interfacial Electrochemistry*, 2nd ed.,
DOI 10.1007/978-3-642-04937-8_3, © Springer-Verlag Berlin Heidelberg 2010

together by hydrogen bonds that are deformed or partially broken, and which are constantly rearranging. Because of their structure – two positively charged hydrogen atoms and two pairs of non-bonding electrons – water molecules have the tendency to form tetrahedral local structures. Figure 3.1 shows a typical snapshot of a few water molecules taken from a computer simulation. The tendency to form hydrogen bonds is evident, even though the bond lengths and angles do not correspond exactly to the hydrogen bonds in ice.

While such snapshots help in visualizing the structure, they do not contain quantitative informations. For this purposes one introduces the *pair correlation function* $g(r)$, which is defined in the following way: Place one molecule at the center of the coordinate system; the density $\rho(r)$ of the other molecules then depends on the distance r. $g(r)$ is obtained by normalizing $\rho(r)$ with respect to the bulk density ρ_0:

$$g(r) = \rho(r)/\rho_0 \tag{3.1}$$

For large separations the density tends towards the bulk value; on the other hand, not two molecules can occupy the same position in space. Therefore:

$$\lim_{r \to 0} g(r) = 0 \qquad \lim_{r \to \infty} g(r) = 1 \tag{3.2}$$

For water the position of the molecule is usually identified with the position of the oxygen atom, which is larger and heavier than the hydrogen atoms. Also, the pair correlation functions of the oxygen atoms can be measured by neutron scattering. The corresponding pair correlation function is shown in Fig. 3.2. The limiting relations of Eq. (3.2) are clearly seen. The nearest neighbors sit at a distance of about 2.7 Å, and give rise to a corresponding maximum. At larger distances minor maxima can be seen, but their height decreases with increasing distance. The structure is loosened at higher temperatures, and the maxima in $g(r)$ are less pronounced.

The pronounced first maximum in the correlation function $g(r)$ is mainly a packing effect. Similar peaks can be obtained in the simplest model for a

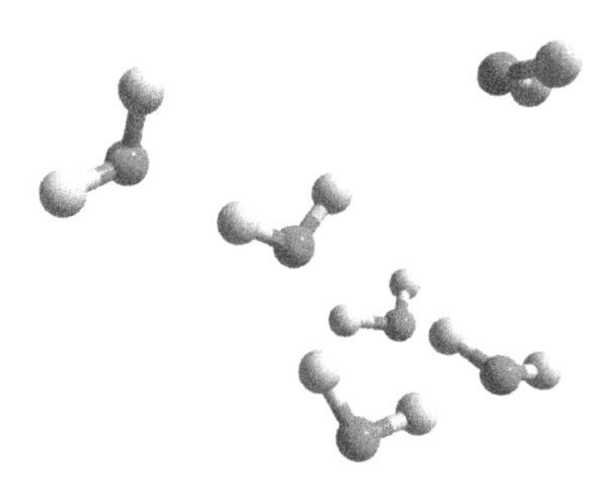

Fig. 3.1. Snapshot of a few neighboring molecules in liquid water, taken from a computer simulation.

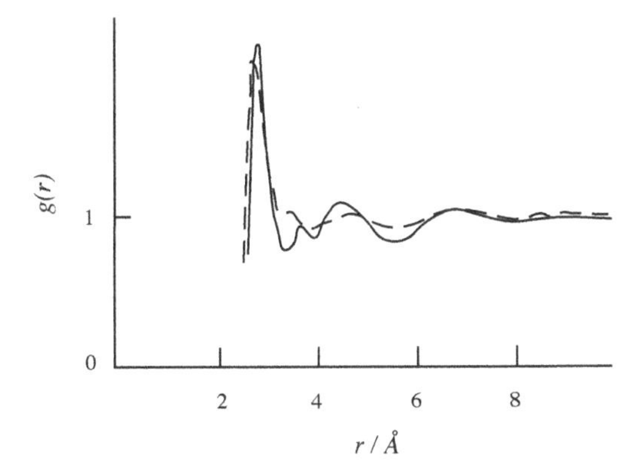

Fig. 3.2. Pair correlation function of water at two different temperatures; the *full line* is for 4°C, the *dashed line* for 75°C. The data have been taken from [2].

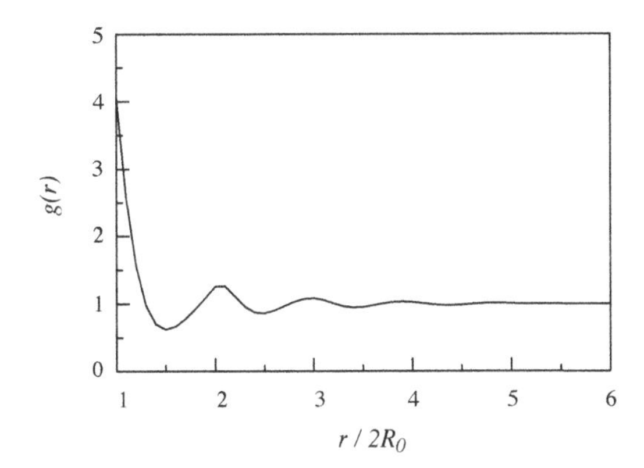

Fig. 3.3. Pair correlation function of an ensemble of hard spheres with a density of $3.2 \times 10^{22}\mathrm{cm}^{-3}$ at ambient temperatures.

liquid, an ensemble of hard spheres. Like the name suggests it consists of identical spheres with a radius R_0, which cannot overlap, but do not interact otherwise. Thus, their interaction potential can be written in the form:

$$V(r) = \begin{cases} \infty & \text{for} \quad r < 2R_0 \\ 0 & \text{for} \quad r \geq 2R_0 \end{cases} \tag{3.3}$$

At sufficiently high densities, the pair correlation function for an ensemble of hard spheres also shows oscillations, whose amplitudes decrease with growing separation (see Fig. 3.3). Of course, they show fewer details than those of water, because specific interactions such as hydrogen bonds are missing.

Pair correlation functions say nothing about the angular distribution or about hydrogen bonding. But even though water has been the subject of innumerable investigations, both experimental and theoretical, the details of its structure are still the subject of debate. The available experimental data leave room for interpretation, and the theoretical treatments use approximations whose validity is difficult to assess. At present, the best model calculations are based on density functional theory, but even this rests on approximations which may work badly for systems with a high electronic polarizability such as water. We refer the interested reader to the cited literature [3].

Electrochemists are especially interested in the structure of water at the electrode surface, where the distribution function $g(x)$ plays the same role that the pair correlation function plays in the bulk. It is defined as the normalized density as a function of the distance x from the electrode surface. There are no experimental data for $g(x)$, and it is difficult to see how they could be obtained. Therefore our knowledge derives from models, which have been implemented in computer simulations. There are several competing semi-empirical models for water, all of which give a good representation of the main features of bulk water. When such models are used to investigate the structure of water at uncharged metal surfaces, they give distribution functions such as those shown in Fig. 3.4, which are characterized by a major peak at the surface, and minor maxima at a larger distances. The orientation of the water molecules in the first layer is mainly such that their dipole moments are parallel to the surface. However, most calculations predict a small net orientation such that on an average the oxygen end is a little closer to the metal than the hydrogens, an effect caused by the interaction of the two lone electron pairs of water with the metal. This leads to a rise of the electrode potential from the surface towards the bulk of the solution, which has the nature of a surface potential as discussed in Chap. 1.

The investigation of water in contact with a metal surface is much easier in ultrahigh vacuum and at low temperatures. Typically, the adsorption of water molecules is rather weak, with energies of the order of 0.3–0.5 eV, and therefore water tends to desorb well below room temperature. The adsorption energy is often lower than the interaction between water molecules, therefore on most metals water forms a hexagonal structure known as the *water bilayer* (see Fig. 3.5), which is accommodated so as to be in registry with the metal lattice. The basic unit is a ring of six water molecules: Three molecules are bonded to the metal through their oxygen ends; their hydrogen atoms are directed towards a secondary layer of three water molecules, which is situated a little further away from the surface. Thus, in the hexagon three molecules are bonded to the surface, the other three are held by hydrogen bonds that connect them to the first layer. Two different conformations exist: In one conformation (H-up), one of the the hydrogen atoms of the second layer is directed away from the surface; in the other (H-down), these atoms point towards the surface. Whether vestiges of these structures exist in electrochemical systems, where

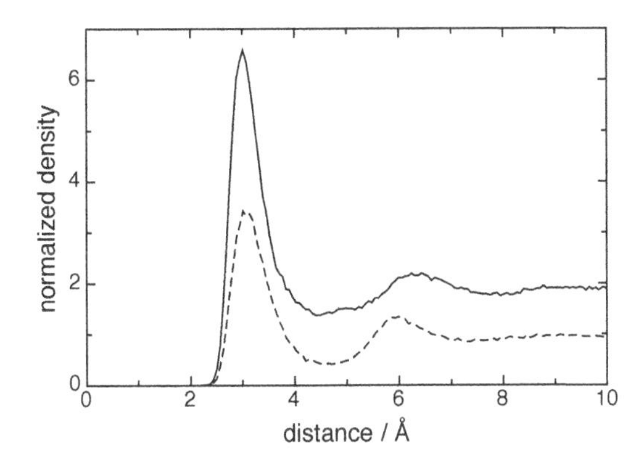

Fig. 3.4. Densities of the oxygen (*dashed line*) and hydrogen atoms (*full line*) of water adsorbed on the surface of a single crystal Ag(111) electrode. The densities have been normalized to unity for oxygen and two for hydrogen.

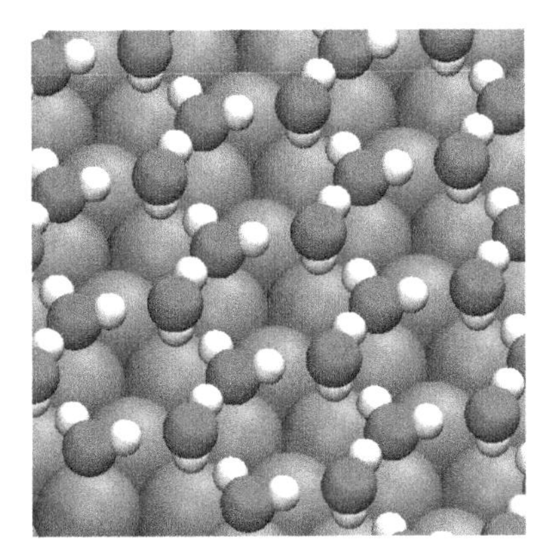

Fig. 3.5. H-down bilayer of water on the surface of Pt(111).

the temperatures are much higher and the surface is in contact with a bulk phase, is an open question.

3.2 Solvatisation of ions

The dipole moments of the solvent molecules induce a strong interaction with ions which is known as solvation or, in the case of water, as hydration. This interaction can be characterized by the free energy of solvation, a term which comprises two related, but different quantities, one that can be measured, and one that cannot. The measurable quantity is the *real free energy of solvation* and is defined as:

Ion	Radius/Å	$-\Delta G_{sol}$	$-\Delta G^r_{sol}$	$-\Delta G^{Born}_{sol}$
Li^+	0.60	5.38	5.32	11.85
Na^+	0.95	4.28	4.28	7.49
K^+	1.33	3.52	3.51	5.35
Rb^+	1.48	3.34	3.29	4.81
Cs^+	1.69	3.09	2.95	4.65
F^-	1.36	4.52	4.32	5.23
Cl^-	1.81	3.30	3.08	3.93
Br^-	1.95	3.16	2.83	3.65
I^-	2.16	2.68	2.49	3.29

Table 3.2. Free hydration enthalpies of ions in eV. The values for ΔG^r_{sol} have been taken from Randles [4], the values for ΔG_{sol} from Latimer [5]; for the latter values the free energy of solvation of the proton has been taken as -11.25 eV. The last column gives the values obtained from the Born formula.

The real free energy ΔG^r_{sol} of solvation is the work that must be expended to transfer an ion from the vacuum into the solution.

It has a negative sign when free energy is gained, which is always the case if the ion is solvable. This definition resembles that for the work function of a metal; thus the free energy of solvation can be considered as the work function of the ion, but with the opposite sign. Conceptually, it can be decomposed into two parts: a bulk part caused by the interaction with the solvent, and a surface part $ze_0\chi$, which is the work expended against the surface dipole potential χ. Table 3.2 gives the free energies of hydration of several simple ions. However, most tables only give the part of the solvation free energy that pertains to the interaction with the solvent and is simply denoted as the *free energy of solvation* ΔG_{sol}. This cannot be measured for a single ion but only for a salt, for which the surface terms cancel. A convention is needed to split the values for a salt into two terms, and this is usually based on the solvation energy of one particular ion such as the proton, which is obtained from a specific model. The value for one ion suffices because the free energies of solvation are additive in the following sense: In a series of salts of the type AX the difference in the free energy A_1X_1 and A_2X_1 equals that between the salts A_1X_2 und A_2X_2, and this is then the difference in the free energies of solvation between A_1 and A_2. This principle holds well at low concentrations of the ions, and in the absence of ion-pair formation. Table 3.2 gives a few of such conventional free energies of hydration.

Just like the inner and the outer potential of a phase, the two different concepts for the free energy of solvation differ only by a surface term:

$$\Delta G^r_{sol} - \Delta G_{sol} = ze_0\chi \tag{3.4}$$

For a given solvent, this difference should only depend on the charge number z. However, a comparison between the corresponding values in Table 3.2 shows that the experimental data are not sufficiently exact to fulfill this relation.

In a simple model due to Born the interaction between an ion and the solvent is considered to be entirely electrostatic; consequently the ion is represented as a charged sphere, the solvent as a continuum with dielectric constant ϵ. The free energy of solvation is then the energy gained when the charged sphere is transferred from the vacuum to the solution. In order to calculate this quantity, we require an expression for the energy of a charged sphere in a dielectric continuum, which equals the energy required to charge a sphere that is initially uncharged.

For this purpose we consider a sphere with radius R and carrying a charge Q embedded in a dielectric. The sphere generates a potential $V(r) = Q/4\pi\epsilon\epsilon_0 r$. Therefore the work:

$$\delta W = \frac{Q}{4\pi\epsilon\epsilon_0 R}\delta Q \tag{3.5}$$

is required to add a small charge δQ to the sphere. The total work to charge the sphere is obtained by integration:

$$W = \int_0^\infty \frac{Q}{4\pi\epsilon\epsilon_0 R}\, dQ = \frac{Q^2}{8\pi\epsilon\epsilon_0 R} \tag{3.6}$$

In order to estimate the free energy of solvation, we subtract the electrostatic energy for the sphere in a dielectric from the corresponding value in the vacuum; the latter is obtained by setting $\epsilon = 1$. This results in the Born formula:

$$\Delta G_{\text{sol}} = -\left(1 - \frac{1}{\epsilon}\right)\frac{(ze_0)^2}{8\pi\epsilon_0 R} \tag{3.7}$$

In this formula, the solvation energy scales with the square of the charge and is hence independent of its sign. As is to be expected, it decreases with increasing ionic radius. Somewhat surprisingly, it depends only weakly on the dielectric constant ϵ. For good solvents, the term $1/\epsilon$ is of the order of 10^{-2} and hence almost negligible compared to unity. So, in this model an ion looses almost all its electrostatic energy when it is transferred from the vacuum into a good solvent.

Naturally, such a simple model can only give the order of magnitude of the solvation energy, and reproduce the main trends (see Table 3.2). Especially for small ions the Born formula overestimates the solvation energy, because in this case the electric field near the ion is large and induces dielectric saturation. In better models, at least the primary solvation shell, which consists of the solvent molecules in contact with the ion, is treated separately. For a further discussion, we refer to [6].

Of all ions, the proton has the highest energy of hydration. It interacts so strongly with water, that it never exists in naked form. The most important complexes are the Zundel $H_5O_2^+$ ion, in which two water molecules share an excess proton, and the Eigen ion $H_9O_4^+$, which is a H_3O^+ with its primary solvation shell (see Fig. 3.6). Both forms have similar energies. Neither complex

Fig. 3.6. Zundel (*left*) and Eigen (*right*) cations.

is stable in a chemical sense; rather, the configuration surrounding the proton is a constantly changing network of hydrogen-bonded water.

The apparent diffusion of protons in water is faster than that of other ions, since a water molecule can accept a proton at one side and pass a proton to the next water molecule at the other side [3]. This so called *Grotthuss mechanism* is schematically depicted in Fig. 3.7. A more complete description takes account of the surrounding water, which participates in the process. Thus, at first the excess proton is localized on the hydronium ion (1), which is the center of an Eigen ion, whose solvation shell is not shown. Then it forms a hydrogen bond with molecule (2), and a Zundel ion results. Through a thermal fluctuation the hydrogen bond is broken again, with the excess proton now residing on molecule (2). The latter then bonds with molecule (3) to form another Zundel ion, which later separates with the proton localized on molecule (3).

Problems

1. An electric field contains an energy; the energy per unit volume is given by:

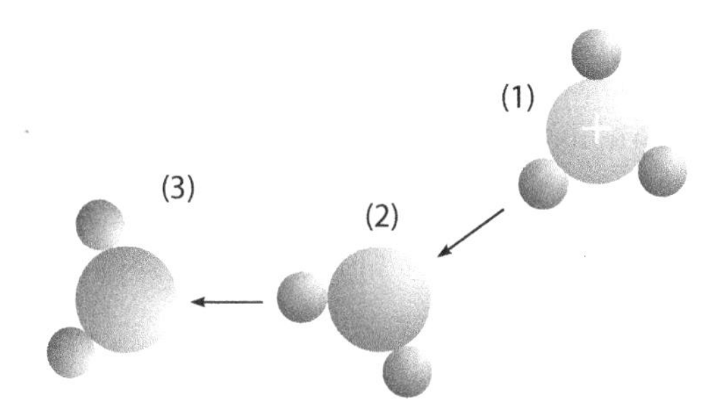

Fig. 3.7. Grotthuss mechanism of proton transfer in water.

$$u = \frac{\epsilon\epsilon_0}{2} E^2 \tag{3.8}$$

where E is the electric field. Derive Eq. (3.7) for the energy of solvation of a charged sphere by integrating the energy density over the space outside of the sphere.

2. Consider a cation with unit charge and radius 1 Å. Calculate its energy of interaction with a point dipole of magnitude 6.14×10^{-30}Cm, sitting at a distance of 2.5 Å and oriented with its negative end towards the ion. Compare this energy with the hydration energy of this ion according to the Born formula.

References

1. Y. Zubavicus and M. Grunze, *Science* **304** (2004) 974; S. Izvekov and G.A. Voth, *J. Chem. Phys.* **122** (2005) 014515 (recent articles about water).
2. A.H. Narten, M.D. Denford, and H.A. Levy, *Disc. Faraday Soc.* **43** (1967) 97.
3. D. Marx, *ChemPhysChem* **7** (2006) 1833.
4. J.E.B. Randles, *Trans. Faraday Soc.* **52** (1956) 1573.
5. W.M. Latimer, *The Oxidation States of the Elements and Their Potentials in Aqueous Solutions.* Prentice Hall, Englewood Cliffs, NJ, 1952.
6. J. O'M. Bockris and A.K.N. Reddy, *Modern Electrochemistry*, Vol. 1, 2nd edition. Springer, New York, NY, 1998.

4

A few basic concepts

In this chapter we introduce and discuss a number of concepts that are commonly used in the electrochemical literature and in the remainder of this book. In particular we will illuminate the relation of electrochemical concepts to those used in related disciplines. Electrochemistry has much in common with *surface science*, which is the study of solid surfaces in contact with a gas phase or, more commonly, with ultrahigh vacuum (uhv). A number of surface science techniques has been applied to electrochemical interfaces with great success. Conversely, surface scientists have become attracted to electrochemistry because the electrode charge (or equivalently the potential) is a useful variable, which cannot be well controlled for surfaces in uhv. This has led to a laudable attempt to use similar terminologies for these two related sciences, and to introduce the concepts of the *absolute scale of electrochemical potentials* and the *Fermi level of a redox reaction* into electrochemistry. Unfortunately, there is some confusion of these terms in the literature, even though they are quite simple.

4.1 The electrochemical potential

In ordinary thermodynamics the chemical potential of a species i is defined as:

$$\mu_i = \left(\frac{\partial G}{\partial N_i} \right)_{p,T} \tag{4.1}$$

where G is the Gibbs energy of the phase under consideration, p denotes the pressure, T the temperature, and N_i the number of particles of species i. So the chemical potential is the work required to add a particle to the system at constant pressure and temperature. Alternatively, one may define μ_i by taking the derivative with respect to m_i, the number of moles of species i. The two definitions differ by a multiplicative constant, Avogadro's constant; we shall use the former definition.

W. Schmickler, E. Santos, *Interfacial Electrochemistry*, 2nd ed.,
DOI 10.1007/978-3-642-04937-8_4,

If the particles of species i in Eq. (4.1) are charged, one speaks of an *electrochemical potential* instead, and writes $\tilde{\mu}_i$. The usual thermodynamic equilibrium conditions are now in terms of the $\tilde{\mu}_i$. For example, if a species i is present both in a phase α and in a phase β, and the interface between α and β is permeable to i, then $\tilde{\mu}_{i,\alpha} = \tilde{\mu}_{i,\beta}$ at equilibrium.

In adding a charged particle work is done against the inner potential ϕ, and it may be useful to separate this out and write:

$$\tilde{\mu}_i = \left(\frac{\partial G}{\partial N_i} \right)_{p,T} = \mu_i + z_i e_0 \phi \tag{4.2}$$

where z_i is the charge number of species i, e_0 is the unit of charge, and μ_i is again called the *chemical potential* since it contains the work done against chemical interactions. For an uncharged species chemical and electrochemical potential are the same.

At zero temperature the electrons in a solid occupy the lowest energy levels compatible with the Pauli exclusion principle. As mentioned in Chap. 2, the highest energy level occupied at $T = 0$ is the Fermi level, E_F. For metals the Fermi level and the electrochemical potential are identical at $T = 0$, since any electron that is added to the system must occupy the Fermi level. At finite temperatures E_F and the electrochemical potential $\tilde{\mu}$ of the electrons differ by terms of the order of $(kT)^2$, which are typically a fraction of a percent and are hence negligible for most purposes. Numerical values of E_F or $\tilde{\mu}$ must refer to a reference point, or energy zero. Common choices are a band edge or the vacuum level, i.e. a reference point in the vacuum at infinity. Obviously, one has to be consistent in the choice of the reference point when comparing the Fermi levels of different systems.

For electrons in a metal the *work function* Φ is defined as the minimum work required to take an electron from inside the metal to a place just outside (c.f. the preceding definition of the outer potential). In taking the electron across the metal surface, work is done against the surface dipole potential χ. So the work function contains a surface term, and it may hence be different for different surfaces of a single crystal. The work function is the negative of the Fermi level, provided the reference point for the latter is chosen just outside the metal surface. If the reference point for the Fermi level is taken to be the vacuum level at infinity instead, then $E_F = -\Phi - e_0\psi$, since an extra work $-e_0\psi$ is required to take the electron from the vacuum level to the surface of the metal. The relations of the electrochemical potential to the work function and the Fermi level are important because one may want to relate electrochemical and solid-state properties.

4.2 Absolute electrode potential

The standard electrode potential [1] of an electrochemical reaction is commonly measured with respect to the standard hydrogen electrode (SHE) [2],

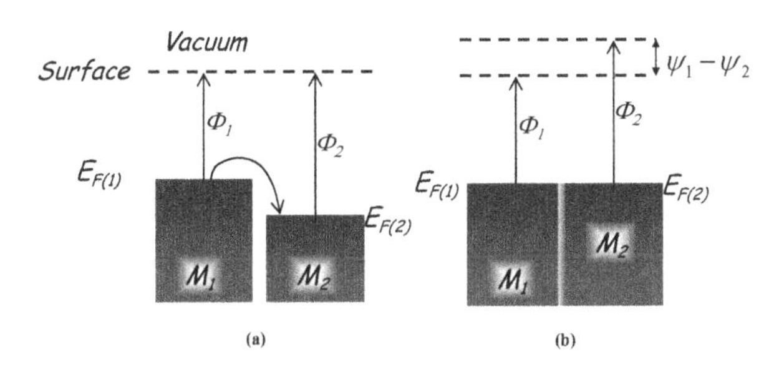

Fig. 4.1. Two metals of different work functions before (**a**) and after (**b**) contact (schematic).

and the corresponding values have been compiled in tables. The choice of this reference is completely arbitrary, and it is natural to look for an absolute standard such as the vacuum level, which is commonly used in other branches of physics and chemistry. To see how this can be done, let us first consider two metals, I and II, of different chemical composition and different work functions Φ_{I} and Φ_{II}. When the two metals are brought into contact, their Fermi levels must become equal. Hence electrons flow from the metal with the lower work function to that with the higher one, so that a small dipole layer is established at the contact, which gives rise to a difference in the outer potentials of the two phases (see Fig. 4.1). No work is required to transfer an electron from metal I to metal II, since the two systems are in equilibrium. This enables us calculate the outer potential difference between the two metals in the following way. We first take an electron from the Fermi level E_{F} of metal I to a point in the vacuum just outside metal I. The work required for this is the work function Φ_{I} of metal I. We then take the electron in the vacuum to a point just above metal II; this requires the work $-e_0(\psi_{\mathrm{II}} - \psi_{\mathrm{I}})$. We then take the electron to the Fermi level of metal II, and gain the energy $-\Phi_{\mathrm{II}}$. Since the total work for this process must be zero, we obtain:

$$\psi_{\mathrm{I}} - \psi_{\mathrm{II}} = \frac{-(\Phi_{\mathrm{I}} - \Phi_{\mathrm{II}})}{e_0} \tag{4.3}$$

so that the outer potential difference can be calculated from the metal work function. By the same reasoning different faces of a single metal crystal have different outer potentials if their work functions are not equal.

We should like to define a "work function" of an electrochemical reaction which enables us to calculate outer potential differences in the same way for a metal-solution interface, and this work function should also refer to the vacuum. For this purpose we consider a solution containing equal amounts of Fe^{3+} and Fe^{2+} ions in contact with a metal M, and suppose that the reaction is at equilibrium. We now transfer an electron from the solution via the vacuum to the metal in the following way:

1. Take an Fe^{2+} ion from the solution into the vacuum above the solution; the work required is the negative of $\Delta G^r_{sol}(Fe^{2+})$, the *real Gibbs energy of solvation* of the Fe^{2+} ion. Real Gibbs energies of solvation are measurable; they include the work done against the surface potential of the solution.
2. Take an electron from the Fe^{2+} ion: $Fe^{2+} \rightarrow Fe^{3+} + e^-$; the work required is the third ionization energy I_3 of Fe.
3. Put the Fe^{3+} back into the solution, and gain $\Delta G^r_{sol}(Fe^{3+})$.
4. Take the electron from just outside the solution across to a position just outside the metal; the work required is $-e_0(\psi_m - \psi_s)$; the index m denotes the metal, s the solution.
5. Take the electron to the Fermi level of the metal, and gain $-\Phi_m$ in energy.

Adding up all the energies, we obtain:

$$-\Delta G^r_{sol}(Fe^{2+}) + I_3 + \Delta G^r_{sol}(Fe^{3+}) - e_0(\psi_m - \psi_s) - \Phi_m = 0 \tag{4.4}$$

or

$$e_0(\psi_m - \psi_s) = \left[\Delta G^r_{sol}(Fe^{3+}) - \Delta G^r_{sol}(Fe^{2+}) + I_3\right] - \Phi_m \tag{4.5}$$

Comparison with Eq. (4.3) suggests that we identify the expression in the square brackets, which depends only on the properties of the redox couple Fe^{3+}/Fe^{2+} in the solution, with the work function of this couple and define:

$$\Phi(Fe^{3+}/Fe^{2+}) = \Delta G^r_{sol}(Fe^{3+}) - \Delta G^r_{sol}(Fe^{2+}) + I_3 \tag{4.6}$$

All the quantities on the right-hand side of this equation are measurable; so this work function is well defined. Fortunately, it is not necessary to calculate the work function for every electrode reaction: The difference between the work functions of two electrode reactions (measured in eV) equals the difference between their standard potentials on the conventional hydrogen scale (measured in V) – this can be easily seen by constructing electrochemical cells with the SHE (standard hydrogen electrode) as a counter electrode. So it is sufficient to know the work function of one particular reaction in a given solvent. For the SHE (i.e. the couple H_2/H^+), the work function is currently estimated as 4.5 ± 0.2 eV; so one obtains the work function of any electrochemical reaction by simply adding this number to the standard potential (in volts) on the SHE scale. By dividing the resulting scale of work functions by the unit charge (or expressing quantities in volts instead of electron volts) one obtains the *absolute scale of electrochemical potentials*.

Since the absolute and the conventional electrode potentials differ only by an additive constant, the absolute potential depends on the concentration of the reactants through the familiar Nernst's equation. This dependence is implicitly contained in Eq. (4.6); the real Gibbs energies of solvation contain an entropic term, which depends on the concentration of the species in the solution.

For a metal, the negative of the work function gives the position of the Fermi level with respect to the vacuum outside the metal. Similarly, the negative of the work function of an electrochemical reaction is referred to as the

Fermi level E_F(redox) of this reaction, measured with respect to the vacuum; in this context *Fermi level* is used as a synonym for electrochemical potential. If the same reference point is used for the metal and the redox couple, the equilibrium condition for the redox reaction is simply: E_F(metal)= E_F(redox). So the notion of a Fermi level for a redox couple is a convenient concept; however, this terminology does not imply that there are free electrons in the solution which obey Fermi–Dirac statistics, a misconception sometimes found in the literature.

The scale of electrochemical work functions makes it possible to calculate the outer potential difference between a solution and any electrode provided the respective reaction is in equilibrium. A knowledge of this difference is often important in the design of electrochemical systems, for example, for electrochemical solar cells. However, in most situations one needs only relative energies and potentials, and the conventional hydrogen scale suffices.

4.3 Three-electrode configuration

Generally electrochemists want to investigate one particular interface between an electrode and an electrolyte. However, to pass a current through the system at least two electrodes are needed. Further, one needs a reference electrode to determine the potential of the working electrode. Since the potential of the reference electrode must remain constant, no current should flow through it. So in practice one takes three electrodes: the working electrode, which one wants to investigate, a counter electrode, which takes up the current, and a reference electrode (see Fig. 4.2). The potential of the working electrode is then measured with respect to that of the reference electrode. It is important that the ohmic potential drop between the working and the reference electrode is as small as possible. One procedure is to keep the reference electrode in a separate compartment, and link it to the main cell with a so-called *Luggin capillary*, whose tip is placed very close to the working electrode. Since no current passes between the working and the reference electrode, the ohmic drop between the two is limited to the region between the capillary tip and the working electrode. There is an additional problem caused by the *junction potential* at the Luggin capillary; a small potential drop is established in the region where two electrolytes of different composition meet [3]. However, in practice these junction potentials can be kept very small and, more importantly, constant, and can be disregarded.

What is actually measured as electrode potential in such a configuration? Consider a metal electrode (M) in equilibrium with a solution containing a redox couple red/ox with a standard hydrogen electrode attached. One measures the electrode potential by taking the two leads of a voltmeter and attaching one to the working and the other to the reference electrode. The latter is made of platinum, and to avoid unnecessary complications we assume that the two leads of the voltmeter are also made of platinum. According to Ohm's

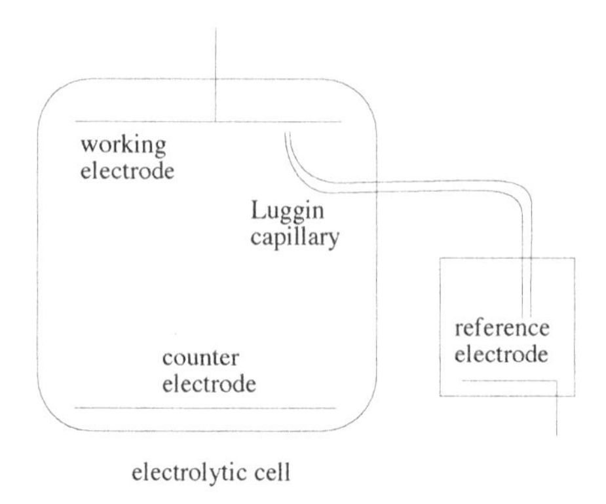

Fig. 4.2. Electrochemical cell with a three-electrode configuration.

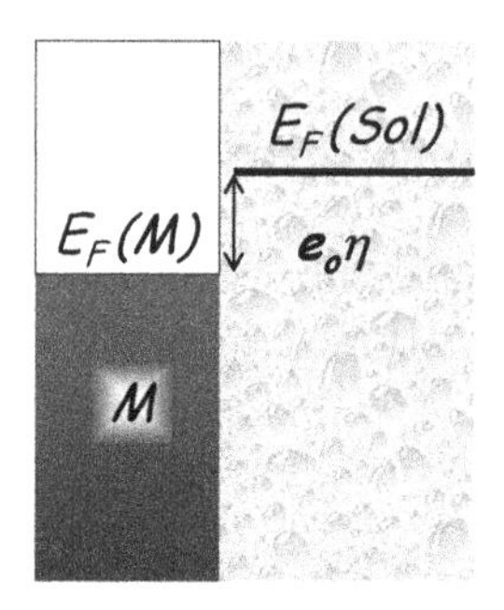

Fig. 4.3. Shift of the metal Fermi level on application of an overpotential.

law the current is proportional to the difference in the driving force, which is the difference in the electrochemical potential. So the voltmeter connected to two phases measures the difference in the electrochemical potential; hence the measured voltage ΔV is given by:

$$-e_0\,\Delta V = \tilde{\mu}_1 - \tilde{\mu}_2 = \mu_1 - e_0\phi_1 - \mu_2 + e_0\phi_2 \tag{4.7}$$

When the two phases have the same chemical composition, the chemical potentials are equal, and then $\Delta V = \phi_1 - \phi_2$, which was already pointed out in Sect. 4.1. In our case both leads are made of the same material, platinum; so the measured electrode potential, which is the equilibrium potential ϕ_0 of the redox couple, is:

$$\phi_0 = \phi_{\rm I} - \phi_{\rm II} = (\phi_{\rm I} - \phi_M) + (\phi_M - \phi_{\rm sol}) + (\phi_{\rm sol} - \phi_{\rm II}) \tag{4.8}$$

Generally, when two phases are in electronic equilibrium, $e_0(\phi_1 - \phi_2) = \mu_1 - \mu_2$. In our case, the wire I is in equilibrium with the metal M, the latter is in equilibrium with the redox couple, and the platinum electrode II is in equilibrium with the reference couple (index "ref"). So we can rewrite Eq. (4.8) as:

$$e_0\phi_0 = (\mu_{\mathrm{I}} - \mu_M) + (\mu_M - \mu_{\mathrm{redox}}) + (\mu_{\mathrm{ref}} - \mu_{\mathrm{II}}) = -\mu_{\mathrm{redox}} + \mu_{\mathrm{ref}} \qquad (4.9)$$

Since the redox couple and the reference system experience the same inner potential ϕ_{sol}, we have:

$$e_0\phi_0 = -\tilde{\mu}_{\mathrm{redox}} + \tilde{\mu}_{\mathrm{ref}} = \Phi_{\mathrm{redox}} - \Phi_{\mathrm{ref}} \qquad (4.10)$$

since the work function is the negative of the electrochemical potential. So one actually measures the difference in the work functions between the redox couple and the reference electrode, and this is independent of the electrode material for a redox couple not involving a reaction with the electrode M.

In the preceding derivation we presumed that equilibrium prevails, so that the Fermi levels of the metal and of the redox couple are equal. This equilibrium can be disturbed by the application of an external electrode potential $\phi \neq \phi_0$, which lowers the electronic energies in the metal, and in particular the Fermi level, by an amount $-e_0\eta$, where $\eta = \phi - \phi_0$ is called the *overpotential* (see Fig. 4.3). Thus the application of an overpotential leads to a difference $-e_0\eta$ in the Fermi levels of the metal and the solution. However, as the equilibrium is disturbed, the reaction proceeds in one direction; current flows and the concentrations of the reactants at the interface will change unless they are kept constant by fast transport processes. Experimental methods for dealing with this difficulty will be discussed in Chaps. 19 and 20. Until then we will generally assume that the concentrations of the reactants are kept constant.

4.4 Surface tension

The correct thermodynamic function to describe the energetics of a system depends on the external conditions. Thus, for a bulk system held at constant temperature and pressure it is the Gibbs energy G, and for constant temperature and volume it is the Helmholtz energy F. Electrochemical interfaces have an extra variable, the electrode potential ϕ. Commonly, they are held at constant temperature, pressure and potential, and are described by the *surface tension* γ, which we will treat in greater detail in Chap. 8. For liquid electrodes, the surface tension can be measured directly as the Gibbs energy required to increase the surface area. For solid electrodes, the absolute value of γ cannot be measured, but, as we shall demonstrate below, changes in γ can.

With the recent advances in computation is has become possible to calculate the energetics of surfaces and interfaces. Leaving the question of the accuracy of such calculations aside, which obviously depends on the complexity of the system considered, we want to relate the quantities calculated to the surface tension. For every type of energy, we can define a surface excess in the following way. We calculate the actual energy of the system, subtract the energy that the system would have if each of the adjoining phases had bulk properties, and divide the result by the area of the surface or interface. Quantum chemical calculations are usually performed at constant temperature, volume, and surface charge density σ on the solid electrode, and thus give the surface Helmholtz energy F_s per unit area. To obtain the surface tension, we perform what is technically known as a Legendre transformation. This is familiar from ordinary thermodynamics: The internal energy U describes a system at constant volume and entropy; by the transformation $F = U - TS$ one obtains the Helmholtz energy, which has temperature and volume as natural variables.

Holding the electrode at constant potential is equivalent to holding the electrons at constant electrochemical potential $\tilde{\mu}_e$. The excess charge σ per area is related to the number of electrons N_e through $\sigma = -e_0(N_e - zN_a)/A$, where N_a is the number of atoms, which is constant, and z their charge number; A is the surface area. Therefore, we obtain the electrochemical potential of the electrons through:

$$-\tilde{\mu}/e_0 = \frac{\partial F_s}{\partial \sigma} \tag{4.11}$$

The surface tension is then defined as:

$$\gamma = F_s + \sigma\tilde{\mu}/e_0 \tag{4.12}$$

and has the electrode potential as its natural variable, since $d\tilde{\mu}/e_0 = -d\phi$. Indeed, keeping all other variables constant, we have:

$$dF_s = -\frac{\tilde{\mu}}{e_0}\,d\sigma \quad d\gamma = dF + \frac{\sigma}{e_0}d\tilde{\mu} + \frac{\tilde{\mu}}{e_0}d\sigma = -\sigma d\phi \tag{4.13}$$

which also shows that the surface tension has an extremum for $\sigma = 0$, the point or potential of zero charge (pzc). Differentiating again gives:

$$\frac{d^2\gamma}{d\phi^2} = -\frac{d\sigma}{d\phi} = -C \tag{4.14}$$

where C is the differential capacity per unit area, which will be treated in more detail in the following chapter. Since the capacity must be positive – otherwise the interface would charge spontaneously – the extremum at the pzc is a maximum. The second part of Eq. (4.13) shows, that changes in the surface tension can be measured by integrating the charge density over the electrode potential.

For liquid electrodes, the equations derived above are exact; on solid electrodes, there is an extra term involving the surface stress. However, this extra term is negligible for all practical purposes, so that the above equations are excellent approximations for solid metals – see Chap. 8 for details.

In the literature there is some confusion concerning the use of the surface tension and the Helmholtz surface energy. In surface science, often F_s/A is called the surface tension. Note that γ and F_s/A agree only at the pzc (see Eq. (4.12)), and the latter does not have a maximum at the pzc, but a positive slope which, because of the relation between $\tilde{\mu}$ and Φ gives the work function. Therefore, using the wrong form of energy can entail, qualitative errors.

Problems

1. Consider a monolayer of water molecules arranged in a square lattice with a lattice constant of 3 Å. The dipole moment of a single molecule is 6.24×10^{-30} Cm. (a) Calculate the potential drop across the monolayer if all dipole moments are parallel and perpendicular to the lattice plane. (b) If the potential drop across the layer is 0.1 V, what is the average angle of the dipole moment with the lattice plane?
2. Following the ideas of Sect. 4.2, devise a suitable cycle to derive the work function of a metal deposition reaction; this will involve the energy of sublimation of the metal.
3. In a simple model for *sp* metals known as *jellium* the ionic charge is smeared out into a constant positive background charge (see also Fig. 1.5). If the metal occupies the region $-\infty < z \leq 0$, the positive charge distribution is given by:

$$n_+(z) = \begin{cases} n_0 \text{ for } z \leq 0 \\ 0 \text{ for } z > 0 \end{cases}$$

In a simple approximation the distribution of the electrons takes the form:

$$n_-(z) = \begin{cases} n_0(1 - A \exp \alpha z) \text{ for } z \leq 0 \\ n_0 B \exp -\alpha z \text{ for } x > 0 \end{cases}$$

Show that for an uncharged metal surface: $A = B = 1/2$, and derive a formula for the surface dipole potential. Cesium has an electronic density of 0.9×10^{22} cm^{-3} and $\alpha \approx 2$ Å^{-1}. Calculate its surface dipole potential.

References

1. A.J. Bard, R. Parsons, and J. Jordan (eds.), *Standard Potentials in Aqueous Solutions.* Dekker, New York, NY, 1985.
2. D.J.G. Ives and G.J. Janz (eds.), *Reference Electrodes.* Academic Press, New York, NY, 1961.
3. K.J. Vetter, *Electrochemical Kinetics.* Academic Press, New York, NY, 1976 (A good treatment of liquid junction potentials is given in).

5

The metal-solution interface

5.1 Ideally polarizable electrodes

The interface between a metal and an electrolyte solution is the most important electrochemical system, and we begin by looking at the simplest case, in which no electrochemical reactions take place. The system we have in mind consists of a metal electrode in contact with a solution containing inert, nonreacting cations and anions. A typical example would be the interface between a silver electrode and an aqueous solution of KF. We further suppose that the electrode potential is kept in a range in which no or only negligible decomposition of the solvent takes place – in the case of an aqueous solution, this means that the electrode potential must be below the oxygen evolution and above the hydrogen evolution region. Such an interface is said to be *ideally polarizable*, a terminology based on thermodynamic thinking. The potential range over which the system is ideally polarizable is known as the *potential window*, since in this range electrochemical processes can be studied without interference by solvent decomposition.

As we pointed out in the introduction, a double layer of equal and opposite charges exists at the interface. In the solution this excess charge is concentrated in a space-charge region, whose extension is the greater the lower the ionic concentration. The presence of this space-charge region entails an excess (positive or negative) of ions in the interfacial region. In this chapter we consider the case in which this excess is solely due to electrostatic interactions; in other words, we assume that there is no specific adsorption. This case is often difficult to realize in practice, but is of principal importance for understanding more complicated situations.

5.2 The Gouy–Chapman theory

A simple but surprisingly good model for the metal-solution interface was developed by Gouy [1] and Chapman [2] as early as 1910. The basic ideas are

W. Schmickler, E. Santos, *Interfacial Electrochemistry*, 2nd ed.,
DOI 10.1007/978-3-642-04937-8_5,

the following: The solution is modeled as point ions embedded in a dielectric continuum representing the solvent; the metal electrode is considered as a perfect conductor. The distribution of the ions near the interface is calculated from electrostatics and statistical mechanics.

To be specific we consider a planar electrode in contact with a solution of a $z - z$ electrolyte (i.e., cations of charge number z and anions of charge number $-z$). We choose our coordinate system such that the electrode surface is situated in the plane at $x = 0$. The inner potential $\phi(x)$ obeys Poisson's equation:

$$\frac{d^2\phi}{dx^2} = -\frac{\rho(x)}{\epsilon\epsilon_0} \tag{5.1}$$

where $\rho(x)$ is the charge density in the electrolyte, ϵ the dielectric constant of the solvent, and ϵ_0 the permittivity of the vacuum. Let $n_+(x)$ and $n_-(x)$ denote the densities of the cations and anions; in the bulk they have the same density n_0. We have:

$$\rho(x) = ze_0 \left[n_+(x) - n_-(x)\right] \tag{5.2}$$

The ionic densities must in turn depend on the potential $\phi(x)$. We choose $\phi(\infty) = 0$ as our reference, and apply Boltzmann statistics:

$$\begin{aligned} n_+(x) &= n_0 \exp -\frac{ze_0\phi(x)}{kT} \\ n_-(x) &= n_0 \exp \frac{ze_0\phi(x)}{kT} \end{aligned} \tag{5.3}$$

Strictly speaking the exponents should not contain the inner potential ϕ but the so-called potential of mean force, but this subtlety is only important at high electrolyte concentrations and high potentials, where other weaknesses of this theory also become important. Substituting Eqs. (5.3) and (5.2) into Eq. (5.1) gives:

$$\frac{d^2\phi}{dx^2} = -\frac{ze_0 n_0}{\epsilon\epsilon_0}\left(\exp -\frac{ze_0\phi(x)}{kT} - \exp \frac{ze_0\phi(x)}{kT}\right) \tag{5.4}$$

which is a nonlinear differential equation for the potential $\phi(x)$ known as the *Poisson–Boltzmann equation.* We first consider the simple case in which $ze_0\phi(x)/kT \ll 1$ everywhere so that the exponentials can be linearized. This gives the *linear Poisson–Boltzmann equation*:

$$\frac{d^2\phi}{dx^2} = \kappa^2\phi(x) \tag{5.5}$$

where κ is the *Debye inverse length*:

$$\kappa = \left(\frac{2(ze_0)^2 n_0}{\epsilon\epsilon_0 kT}\right)^{1/2} \tag{5.6}$$

Concentration/mol l^{-1}	10^{-4}	10^{-3}	10^{-2}	10^{-1}
Debye length/Å	304	96	30.4	9.6

Table 5.1. Debye length for an aqueous solution of a completely dissociated 1–1 electrolyte at room temperature.

$L_D = 1/\kappa$ is the *Debye length*; Table 5.1 shows values for several concentrations of a 1–1 electrolyte in an aqueous solution at room temperature. The solution compatible with the boundary condition $\phi(\infty) = 0$ has the form: $\phi(x) = A \exp(-\kappa x)$, where the constant A is fixed by the charge balance condition:

$$\int_0^\infty \rho(x)\, dx = -\sigma \tag{5.7}$$

where σ is the surface charge density on the metal. $\rho(x)$ is obtained from $\phi(x)$ via Poisson's equation, and a straightforward calculation gives:

$$\phi(x) = \frac{\sigma}{\epsilon\epsilon_0\kappa} \exp(-\kappa x) \tag{5.8}$$

for the potential and:

$$\rho(x) = -\sigma\kappa \exp(-\kappa x) \tag{5.9}$$

for the charge density. So the excess charge on the metal is balanced by a space-charge layer, which decays exponentially in the solution. This configuration of charges obviously has a capacity. The electrode potential is: $\phi = \phi(0) = \sigma/\epsilon\epsilon_0\kappa$ – dipole potentials are ignored in this simple model. The interfacial capacity per unit area, known as the *double-layer capacity*, is:

$$C = \epsilon\epsilon_0\kappa \tag{5.10}$$

So the double-layer capacity is the same as that of a parallel-plate capacitor with the plate separation given by the Debye length. Since for high concentrations the latter is of the order of a few Ångstroms, these capacities can be quite high.

While Eqs. (5.9) and (5.10) are quite instructive, they are valid for small charge densities on the electrode only. For a $z - z$ electrolyte the nonlinear Poisson–Boltzmann equation (5.4) can be solved explicitly. We are mainly interested in the *differential capacity*, defined as $C = \partial\sigma/\partial\phi$, which is a measurable quantity. A short calculation, whose details are given in the appendix of this chapter, gives:

$$C = \epsilon\epsilon_0\kappa \cosh\left(\frac{ze_0\phi(0)}{2kT}\right) \tag{5.11}$$

This is not a useful form since the potential $\phi(0)$ cannot be measured. The electrode potential ϕ differs from $\phi(0)$ by a constant; when $\phi(0) = 0$ the

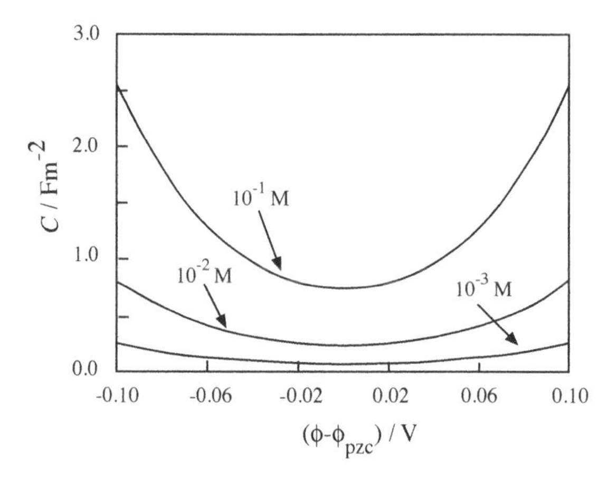

Fig. 5.1. Gouy–Chapman capacity for various concentrations of a 1–1 electrolyte in aqueous solution at room temperature.

electrode carries no charge, and the corresponding electrode potential $\phi_{\rm pzc}$ is the *potential of zero charge* (pzc). So we rewrite Eq. (5.11) in the form:

$$C = \epsilon\epsilon_0\kappa \, \cosh\left(\frac{ze_0\,(\phi - \phi_{\rm pzc})}{2kT}\right) \tag{5.12}$$

This differential capacity is known as the *Gouy–Chapman capacity.* It has a pronounced minimum at the pzc, and it increases with the square root of the electrolyte concentration. Figure 5.1 shows the Gouy–Chapman capacity calculated for several electrolyte concentrations.

Because of the simple model on which it is based, the validity of the Gouy–Chapman theory is limited to low concentrations and small excess charge densities. Even at 5×10^{-2} M solutions, there are substantial deviations at potentials away from zero charge. As an example, Fig. 5.2 shows the capacity of single-crystal silver electrodes in a 5×10^{-2} M solution of a weakly adsorbing electrolyte. All three curves show the characteristic capacity minimum at the pzc, but the deviations from Gouy–Chapman theory away from the pzc are quite evident. The plot also illustrates the dependence of the pzc on the crystal face.

5.3 The Helmholtz capacity

At low electrolyte concentrations, up to about a 10^{-3} M solution, the Gouy–Chapman theory agrees quite well with experimental values of the double layer capacity for nonadsorbing electrolytes. At higher concentrations systematic deviations are observed. In fact the experimental values follow an equation of the form:

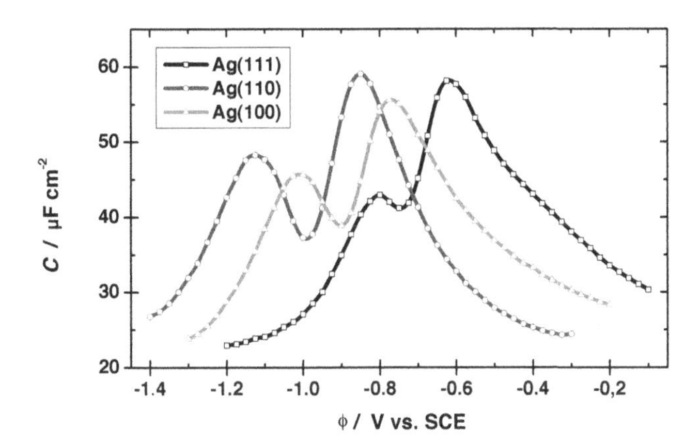

Fig. 5.2. Experimental capacity of single-crystal silver electrodes in a solution of 5×10^{-2} M $KClO_4$.

$$\frac{1}{C} = \frac{1}{C_{GC}} + \frac{1}{C_H} \tag{5.13}$$

where C_{GC} is the Gouy–Chapman capacity given by Eq. (5.12), and the *Helmholtz capacity* C_H is independent of the electrolyte concentration.

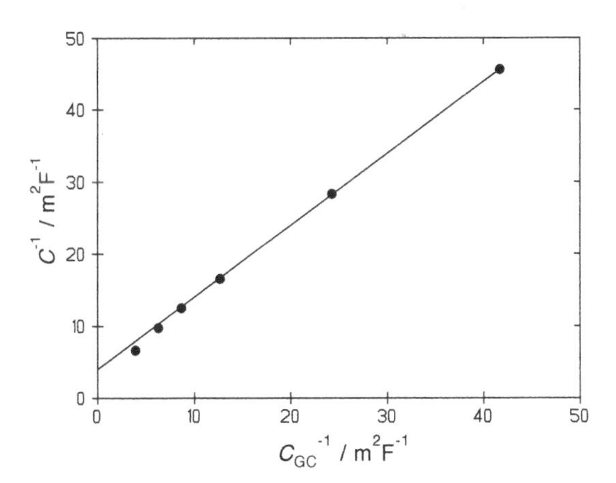

Fig. 5.3. Parsons and Zobel plot; the intercept gives the inverse Helmholtz capacity.

Experimentally the Helmholtz capacity can be obtained by measuring the interfacial capacity C per unit area for several concentrations, and plotting $1/C$ versus the calculated inverse Gouy–Chapman capacity $1/C_{GC}$ at a constant surface charge density σ (*Parsons and Zobel* plot); the intercept of the resulting straight line gives $1/C_H$ (see Fig. 5.3). If the electrode area is not known, one plots the capacity instead and obtains the area from the slope of the plot. If a Parsons and Zobel plot does not result in a straight line, this is an indication that specific adsorption occurs.

The Helmholtz capacity C_H dominates at high electrolyte concentrations, when the extension of the space-charge layer is small, and hence its origin must be in a narrow region right at the interface. For a given system C_H generally depends strongly on the charge density σ and somewhat more weakly on temperature. The capacity-charge characteristics C_H versus σ vary greatly with the nature of the metal and the solvent, and are even somewhat different for different faces of a single crystal. However, they depend only weakly on the nature of the ions in the solution, as long as they are not specifically adsorbed. Figure 5.4 shows capacity-charge characteristics for mercury and for a single crystal silver electrode in contact with an aqueous solution; notice the maximum near the pzc, and how much smaller the capacity of mercury is.

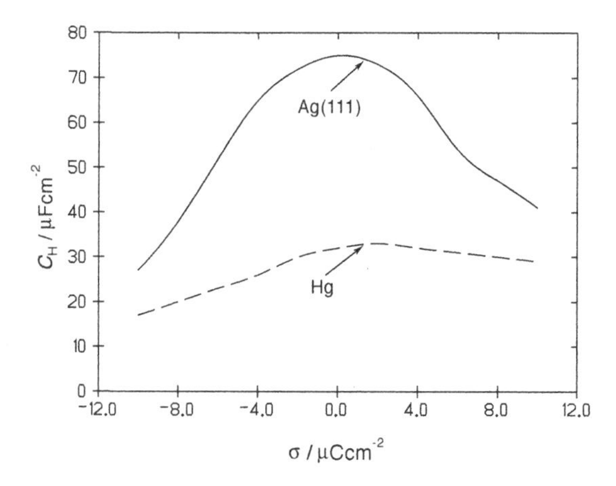

Fig. 5.4. Helmholtz capacity for Ag(111) and mercury in aqueous solutions.

Several theories have been proposed to explain the origin and the order of magnitude of the Helmholtz capacity. Though differing in details, recent theories agree that the Helmholtz capacity contains contributions both from the metal and from the solution at the interface:

1. Due to the finite size of the ions and the solvent molecules, the solution shows considerable structure at the interface, which is not accounted for in the simple Gouy–Chapman theory. The occurrence of a decrease of C from the maximum near the pzc is caused by dielectric saturation, which lowers the dielectric constant and hence the capacity for high surface-charge densities.

2. The surface potential χ of the metal varies with the surface charge. A little thought shows that the change in the surface potential opposes the applied external potential, thus decreasing the total potential drop for a given surface charge and increasing the capacity.

The latter effect can be understood within a simple model for metals: the *jellium model*, which is based on the following ideas: As is generally known, a metal consists of positively charged ions and negatively charged electrons. In the jellium model the ionic charge is smeared out into a constant positive background charge, which drops abruptly to zero at the metal surface. The electrons are modeled as a quantum-mechanical plasma interacting with the background charge and with any external field such as that caused by surface charges. Due to their small mass the electrons can penetrate a little into the solution; typically the electronic density decreases exponentially with a decay length of about 0.5 Å. Since the electronic density of metals is high, this gives rise to an appreciable negative excess charge outside the metal, which for an uncharged surface must be balanced by an equal and opposite positive excess charge within the metal. The resulting electronic charge distribution, plotted as a function of the distance x from the metal surface, is shown in Fig. 5.5; it carries a surface dipole moment which gives rise to a surface potential χ of the order of several volts.

The electric field in the double layer distorts the electronic distribution and changes the surface potential χ. A negative surface charge creates an excess of electrons on the surface. The resulting electrostatic field pulls the electrons toward the solution, and increases the surface dipole potential. Conversely, a

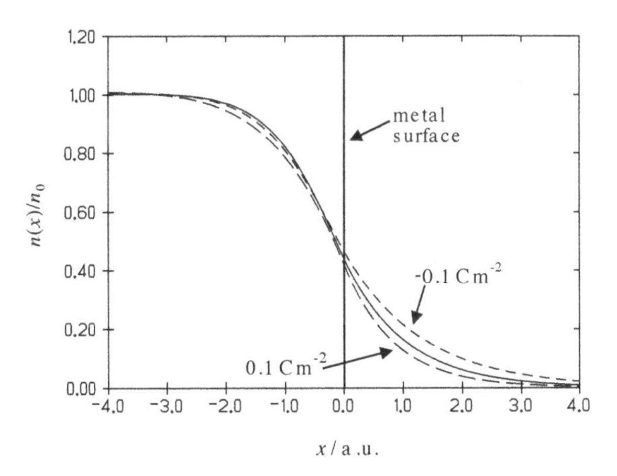

Fig. 5.5. Distribution of the electronic density in the *jellium* model; the metal occupies the region $x \leq 0$. The *unmarked curve* is for an uncharged surface, the other *two curves* are for the indicated surface-charge densities. The distance along the x axis is measured in atomic units (a.u.), where 1 a.u. of length = 0.529 Å.

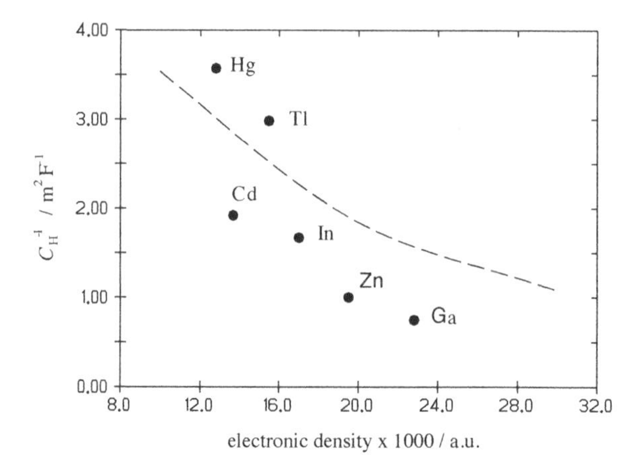

Fig. 5.6. The inverse Helmholtz capacity at the pzc as a function of the electronic density; the latter is plotted in atomic units (a.u.), where 1 a.u. of density $= 6.76 \times 10^{24}$ cm^{-3}. The *dashed line* is based on a model calculation of Schmickler and Henderson [3].

positive excess charge gives rise to a deficiency of electrons, and the surface dipole potential becomes smaller. The change in dipole potential opposes the change in the external potential, and hence increases the capacity. In other words, the electrons at the metal surface form a highly polarizable medium, which enhances the double-layer capacity. Since this is an electronic effect, one might expect that its magnitude increases with the electronic density of the metal. This seems indeed to be the case for simple metals, the *sp* metals of the second and third column of the periodic table (see Fig. 5.6); the Helmholtz capacity of these elements at the pzc correlates with their electronic densities.

5.4 The potential of zero charge

The potential of zero charge (pzc) is a characteristic potential for a given interface, and hence is of obvious interest. In the absence of specific adsorption, it can be measured as the potential at which the Gouy–Chapman capacity obtains its minimum; this value must be independent of the electrolyte concentration, otherwise there is specific adsorption. The pzc coincides with the maximum of the surface tension (see Sect. 4.4), which can be measured directly for liquid metals.

An interesting correlation exists between the work function of a metal and its pzc in a particular solvent. Consider a metal *M* at the pzc in contact with a solution of an inert, nonadsorbing electrolyte containing a standard platinum/hydrogen reference electrode. We connect a platinum wire (label I) to the metal, and label the platinum reference electrode with II. This setup is very similar to that considered in Sect. 4.3, but this time the metal-solution interface is not in electronic equilibrium. The derivation is simplified if we assume that the two platinum wires have the same work function, so that their surface potentials are equal. The electrode potential is then:

$$\phi_{\rm pzc} = \phi_{\rm I} - \phi_{\rm II} = \psi_{\rm I} - \psi_{\rm II} = (\psi_{\rm I} - \psi_M) + (\psi_M - \psi_{\rm sol}) + (\psi_{\rm sol} - \psi_{\rm II}) \quad (5.14)$$

The first and the last term can again be expressed through the work function differences, but not the second term, since this interface is not in electronic equilibrium:

$$\begin{aligned} \phi_{\rm pzc} &= \frac{1}{e_0}\left[(\Phi_M - \Phi_{Pt}) + (\Phi_{Pt} - \Phi_{\rm ref})\right] + (\psi_M - \psi_{\rm sol}) \\ &= \frac{1}{e_0}(\Phi_M - \Phi_{\rm ref}) + (\psi_M - \psi_{\rm sol}) \end{aligned} \quad (5.15)$$

To evaluate the last term we go through a cycle taking a test charge (not an electron!) from outside the metal first into the bulk of the metal, then through the metal-solution interface, then to a position just outside the solution, and finally back to outside the metal. This gives:

$$\psi_M - \psi_{\rm sol} = -\chi_M + \chi_{\rm int} + \chi_{\rm sol} \quad (5.16)$$

where $\chi_{\rm int}$ is the surface potential at the metal-solution interface. If the metal and the solvent did not interact, $\chi_{\rm int}$ would simply be $\chi_M - \chi_{\rm sol}$, and the outer potential difference would vanish at the pzc. However, the metal-solvent interaction modifies the surface potentials; the presence of the solvent changes the distribution of the electrons at the surface, and the interaction of the solvent with the metal surface can lead to a small net orientation of the

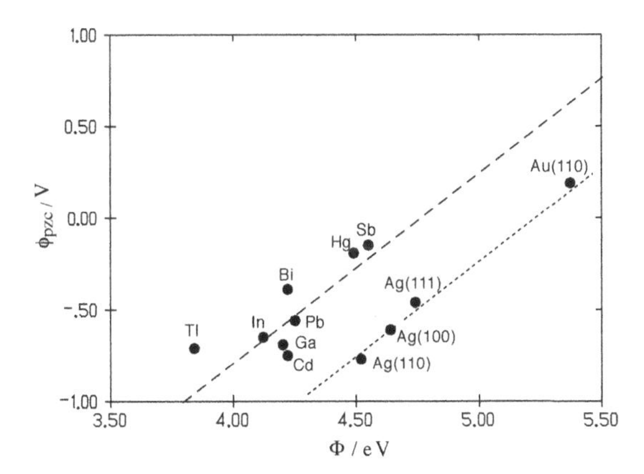

Fig. 5.7. The potential of zero charge (vs. SHE) of metals in aqueous solution; the *upper line* is for *sp* metals, the *lower* for *sd* metals [4].

solvent dipoles. Denoting these changes in the surface potentials by $\delta\chi_M$ and $\delta\chi_{\text{sol}}$, we have: $\psi_M - \psi_{\text{sol}} = \delta\chi_M - \delta\chi_{\text{sol}}$, so that we obtain for the pzc:

$$\phi_{\text{pzc}} = \frac{1}{e_0}(\Phi_M - \Phi_{\text{ref}}) + \delta\chi_M - \delta\chi_{\text{sol}} \tag{5.17}$$

The changes in the dipole potentials are typically small, of the order of a few tenths of a volt, while work functions are of the order of a few volts. If we keep the solvent, and hence Φ_{ref}, fixed and vary the metal, the potential of zero charge will be roughly proportional to the work function of the metal. This is illustrated in Fig. 5.7. A more detailed consideration of the dipole potentials leads to a subdivision into separate correlations for *sp*, *sd*, and transition metals [4].

Problems

1. For a z–z electrolyte define the excess distribution of the cations and anions through: $\delta n_+(x) = n_+(x) - n_0$ and $\delta n_-(x) = n_-(x) - n_0$. Show that $|\delta n_+(x)| = |\delta n_-(x)|$ holds for the linear Gouy–Chapman theory, but not for the nonlinear version.
2. Consider a point dipole with dipole moment m in an external electric field E oriented along the z axis. Choosing a suitable coordinate system, show that the average value of the dipole moment along the direction of the field is:

$$\langle m_z \rangle = \frac{\int_0^{2\pi} d\phi \int_0^{\pi} d\theta \, \sin\theta \; m\cos\theta \; \exp\left(\frac{mE\cos\theta}{kT}\right)}{\int_0^{2\pi} d\phi \int_0^{\pi} \sin\theta \; d\theta \exp\left(\frac{mE\cos\theta}{kT}\right)}$$

For $mE \ll kT$ the exponentials can be expanded. Show that in this limit:

$$\langle m_z \rangle = \frac{m^2 E}{3kT}$$

3. The Thomas–Fermi model of a metal is similar to the Gouy–Chapman theory for electrolytes. In this model the surface-charge density σ is spread over a thin boundary layer. If the metal occupies the region $x \leq 0$, the distribution of the charge density is given by:

$$\rho(x) = A \exp \frac{x}{L_{\text{TF}}}$$

where A is a constant to be determined by charge balance, and L_{TF} is the *Thomas–Fermi length*, which is mainly determined by the electronic density of the metal. Combine this model with the linear Gouy-Chapman theory and derive:

$$\frac{1}{C} = \frac{1}{\epsilon\epsilon_0\kappa} + \frac{L_{\text{TF}}}{\epsilon_0}$$

Compare this result with Eq. (5.13). For most metals $L_{\text{TF}} \approx 0.5$ Å. By examining the experimental data in Fig. 5.6, show that this model cannot explain the origin of the Helmholtz capacity.

Appendix: the nonlinear Gouy–Chapman theory

We rewrite the nonlinear Poisson Boltzmann Eq. (5.4) in the form:

$$\frac{d^2\phi}{dx^2} = -\frac{2ze_0n_0}{\epsilon\epsilon_0}\sinh\frac{ze_0\phi(x)}{kT} \tag{5.18}$$

and multiply both sides by $2d\phi/dx$. Using

$$\frac{d}{dx}\left(\frac{d\phi}{dx}\right)^2 = 2\frac{d^2\phi}{dx^2}\frac{d\phi}{dx} \tag{5.19}$$

we can integrate both sides:

$$\begin{aligned} 2\int_0^\infty \frac{d^2\phi}{dx^2}\frac{d\phi}{dx}\,dx &= \left.\left(\frac{d\phi}{dx}\right)^2\right|_0^\infty \\ &= -\int_0^\infty \frac{4ze_0n_0}{\epsilon\epsilon_0}\sinh\left(\frac{ze_0\phi}{kT}\right)\frac{d\phi}{dx}\,dx \end{aligned} \tag{5.20}$$

Both the field E and the potential ϕ vanish at ∞; so we obtain:

$$E(0)^2 = \frac{4kTn_0}{\epsilon\epsilon_0}\left(\cosh\frac{ze_0\phi(0)}{kT} - 1\right) \tag{5.21}$$

According to Gauss's theorem, $E(0) = \sigma/\epsilon\epsilon_0$; using the identity $\cosh x - 1 = 2\sinh^2 x/2$ gives:

$$\sigma = (8kTn_0\epsilon\epsilon_0)^{1/2}\sinh\frac{ze_0\phi(0)}{2kT} \tag{5.22}$$

Differentiation then gives the Gouy–Chapman expression Eq. (5.11).

Sometimes one requires not only the capacity but the potential $\phi(x)$; we sketch the derivation. If we integrate Eq. (5.18) from x to ∞, we obtain by the same arguments for the derivative $\phi'(x)$:

$$\phi'(x) = -\left(\frac{8kTn_0}{\epsilon\epsilon_0}\right)^{1/2}\sinh\frac{ze_0\phi(x)}{2kT} \tag{5.23}$$

Substituting $\psi(x) = [ze_0\phi(x)]/2kT$ gives:

$$\frac{\psi'(x)}{\sinh\psi(x)} = -\kappa \tag{5.24}$$

where κ is the inverse Debye length. Integration gives:

$$\ln\tanh\frac{\psi}{2} = -\kappa x + \ln C \tag{5.25}$$

where $\ln C$ is the constant of integration, which can be expressed through the value of the potential at the origin:

$$C = \tanh \frac{ze_0\phi(0)}{4kT} \tag{5.26}$$

Equation (5.22) relates $\phi(0)$ to the charge density σ:

$$\frac{ze_0\phi(0)}{2kT} = \text{arcsinh}\ \alpha\sigma, \text{ where } \alpha = (8kTn_0\epsilon\epsilon_0)^{-1/2} \tag{5.27}$$

Using the identity:

$$\tanh\left(\frac{1}{2}\text{arcsinh}\ x\right) = \frac{\sqrt{1+x^2}-1}{x} \tag{5.28}$$

gives finally:

$$\tanh \frac{ze_0\phi(x)}{4kT} = \frac{\sqrt{1+\alpha^2\sigma^2}-1}{\alpha\sigma} \exp -\kappa x \tag{5.29}$$

for the potential.

References

1. G. Gouy, *J. Phys.* **9** (1910) 457.
2. D.L. Chapman, *Philos. Mag.* **25** (1913) 475.
3. W. Schmickler and D. Henderson, *Prog Surf Sci* **22**(4) (1986) 323–420.
4. S. Trasatti, *Advances in Electrochemistry and Electrochemical Engineering*, Vol. 10, edited by H. Gerischer and C.W. Tobias. Wiley Interscience, New York, NY, 1977.

6

Adsorption on metal electrodes: principles

6.1 Adsorption phenomena

Whenever the concentration of a species at the interface is greater than can be accounted for by electrostatic interactions, we speak of *specific adsorption*. It is usually caused by chemical interactions between the adsorbate and the electrode, and is then denoted as *chemisorption*. In some cases adsorption is caused by weaker interactions such as van der Waals forces; we then speak of *physisorption*. Of course, the solvent is always present at the interface; so the interaction of a species with the electrode has to be greater than that of the solvent if it is to be adsorbed on the electrode surface. Adsorption involves at least a partial desolvation. Cations tend to have a firmer solvation sheath than anions, and are therefore less likely to be adsorbed.

The amount of adsorbed species is usually given in terms of the *coverage* θ, which is the fraction of the electrode surface covered with the adsorbate. When the adsorbate can form a complete monolayer, θ equals the ratio of the amount of adsorbate present to the maximum amount that can be adsorbed. In a few systems the area covered by a single adsorbed molecule changes with coverage; for example, some organic molecules lie flat at low coverage and stand up at higher coverages. In this case one must specify to which situation the coverage refers. Another definition of the coverage, often used in surface science, is the following: θ is the ratio of the number of adsorbed species to the number of surface atoms of the substrate. Fortunately, most authors state which definition they use.

The chemisorption of species occurs at specific sites on the electrode, for example on top of certain atoms, or in the bridge position between two atoms. Therefore, most adsorption studies are performed on well-defined surfaces, which means either on the surface of a liquid electrode or on a particular surface plane of a single crystal. Nowadays, most work is done on single crystals, and mercury, which was extensively used before, has almost dropped out of use.

W. Schmickler, E. Santos, *Interfacial Electrochemistry*, 2nd ed.,
DOI 10.1007/978-3-642-04937-8_6, © Springer-Verlag Berlin Heidelberg 2010

Nevertheless, liquid electrodes are not only easier to prepare than single crystal surfaces of solid electrodes, they also have another advantage: Adsorption can be studied by measuring the variation of the surface tension. We defer the thermodynamics of interfaces till Chap. 8; here we merely state that such measurements yield the total *surface excess* Γ_i of a species. Roughly speaking, Γ_i is the amount of species i per unit area in excess over the amount that would be present if its concentration were the same at the interface as in the bulk. Γ_i can be positive or negative – cations, for example, can be excluded from the surface region near a positively charged electrode. Until the advent of modern spectroscopic methods for studying the electrochemical interface, thermodynamic measurements were the only reliable way of determining specific adsorption. While such measurements are easier to perform on liquid electrodes, they have been extended to solid surfaces.

The surface excesses Γ_i not only include the atoms or molecules of species i that are adsorbed on the metal surface, but also those that are in the space-charge region considered in Chap. 5. The latter is also known as the *diffuse part of the double layer*, or simply as the *diffuse double layer*; in contrast, adsorbed particles are said to be part of the *compact part of the double layer*. Usually we are only interested in the amount of particles that is specifically adsorbed. The excess of species i in the diffuse double layer can be minimized by working with a high concentration of nonadsorbing, inert (also called *supporting*) electrolyte. Suppose we want to study the adsorption of an anion A^-; if we add a large excess of nonadsorbing ions B^+ and C^- to the solution, keeping the electrode charge constant, the amount of ions A^- in the diffuse layer will be greatly reduced. This can be seen from the following argument. Let Q be the total surface charge at the interface, that is, Q contains both the excess charge on the metal and the charge of any adsorbed species. This charge must be balanced by a charge $-Q$ in the space-charge region. The concentration $n_i(x)$ of a species i in this region is proportional to its bulk concentration $n_{i,o}$ (see Eq. 5.3): $n_i(x) = n_{i,o} \exp[z_i e_0 \phi(x)/kT]$. Adding an excess of an inert electrolyte will therefore drastically reduce the amount of ions A^- in this region. Since the charge number enters into the exponent, highly charged ions require a higher concentration of supporting electrolyte. In practice it may not be easy to find an inert electrolyte that is not specifically adsorbed; obviously, coadsorption of other ions should be avoided since it drastically changes the conditions at the electrode surface, and makes the interpretation of the experimental data difficult.

6.2 Adsorption isotherms

Consider the adsorption of a species A with concentration c_A in the bulk of the solution. The variation of the coverage θ with c_A, keeping all other variables fixed, is known as the *adsorption isotherm*.

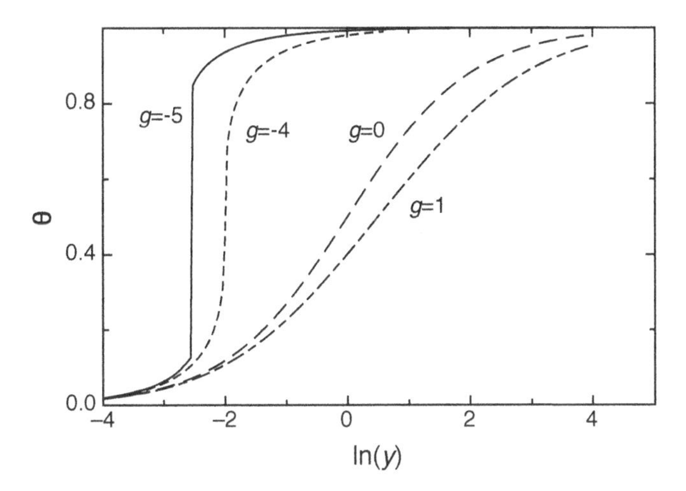

Fig. 6.1. Frumkin isotherms for various values of the adsorbate interaction parameter g; the Langmuir isotherm corresponds to $g = 0$. The abscissa is: $\ln y = \ln(c/c_0) + (\mu_{sol} - \mu_{ad})/kT$.

We first consider the simplest possible model, in which adsorption occurs at fixed sites, and the interaction between adsorbed particles can be neglected. Let the surface contain N adsorption sites, of which M are occupied, and let ϵ_{ad} be the adsorption energy per particle. The internal energy of the adsorbate is simply $M\epsilon_{ad}$. To obtain the free energy, we need the entropy, which according to the Boltzmann formula is: $S = k \ln W$, where W is the number of realizations of the system, or the number of ways of selecting M sites out of N. Therefore:

$$F = M\epsilon_{ad} - kT \ \ln \frac{N!}{M!(N-M)!} \tag{6.1}$$

Using Stirling's formula: $\ln n! \approx n \ln n - n$ for large n gives:

$$F = M\epsilon_{ad} + \left[M \ln \frac{M}{N} + (N-M) \ln \frac{N-M}{N} \right] \tag{6.2}$$

At equilibrium, the chemical potential of the adsorbate must equal the chemical potential of the same particle in the solution. For the adsorbate we obtain:

$$\mu_{ad} = \frac{\partial F}{\partial M} = \epsilon_{ad} + kT \ \ln \frac{\theta}{1-\theta} \tag{6.3}$$

where $\theta = M/N$ is the coverage. The chemical potential for an ideal solute has the form:

$$\mu_{sol} = \mu_0 + kT \ln \frac{c}{c_0} \tag{6.4}$$

where c is the concentration, and the unit concentration c_0 makes the argument of the logarithm dimensionless. Setting the chemical potentials equal results in the *Langmuir isotherm*:

$$\frac{\theta}{1-\theta} = \frac{c}{c_0} \exp\left(\frac{\mu_{\text{sol}} - \mu_{\text{ad}}}{kT}\right) \tag{6.5}$$

The model from which it has been derived is the two-dimensional *lattice-gas model*. We shall meet a three-dimensional version in Chap. 18.

So far, we have ignored interactions between the adsorbates completely. In a simple, phenomenological way one can account for such interactions by assuming that $\Delta\mu_{\text{ad}}$ is proportional to θ: $\mu_{\text{ad}} = \mu_{\text{ad}}^0 + \gamma\theta$, where the constant γ is positive if the adsorbed particles repel, and negative if they attract each other. The resulting isotherm:

$$\frac{\theta}{1-\theta} = \frac{c}{c_0} \exp\left(\frac{\mu_{\text{sol}} - \mu_{\text{ad}}}{kT}\right) e^{-g\theta} \tag{6.6}$$

with $g = \gamma/RT$, is known as the *Frumkin isotherm*. At present there is no general satisfactory theory for adsorbate–adsorbate interaction at electrochemical interfaces, and consequently none for adsorption isotherms. Besides Eqs. (6.5) and (6.6), various other isotherms have been proposed based on rather simple models, but none of them is really satisfactory. Figure 6.1 shows Frumkin isotherms for a few different values of the interaction parameter g. Positive values of g broaden the isotherm because the adsorbed particles repel each other; for negative values of g the isotherms are narrow because adsorption is then a cooperative effect. For $g < -4$, a phase transition occurs, in which the adsorbate condenses on the electrode. This gives rise to a vertical slope in the isotherm. The case $g = 0$ corresponds to the Langmuir isotherm.

The difference $\Delta\mu = \mu_{\text{sol}} - \mu_{\text{ad}}$ depends on the electrode potential ϕ. This dependence will be different for anions, cations, and neutral species. The simplest possible case is the adsorption and total discharge of an ion according to the equation:

$$A^{z+} + ze^- \rightleftharpoons A_{\text{ad}} \tag{6.7}$$

obeying the Langmuir isotherm with a potential dependence of the form:

$$\Delta\mu(\phi) = \Delta\mu(\phi_0) + ze_0(\phi - \phi_0) \tag{6.8}$$

where ϕ_0 is a suitably chosen reference potential. The choice of ϕ_0 is not important since it just determines the zero of the potential scale. A convenient choice may be one for which the coverage is $\theta = 1/2$ for a given electrolyte concentration. The resulting isotherm takes the form:

$$\frac{\theta}{1-\theta} = \frac{c}{c_0} K \exp\left(-\frac{ze_0(\phi - \phi_0)}{kT}\right) \tag{6.9}$$

which is illustrated in Fig. 6.2. However, the assumptions on which this equation is based rarely hold in practice. When ions are adsorbed, the interaction

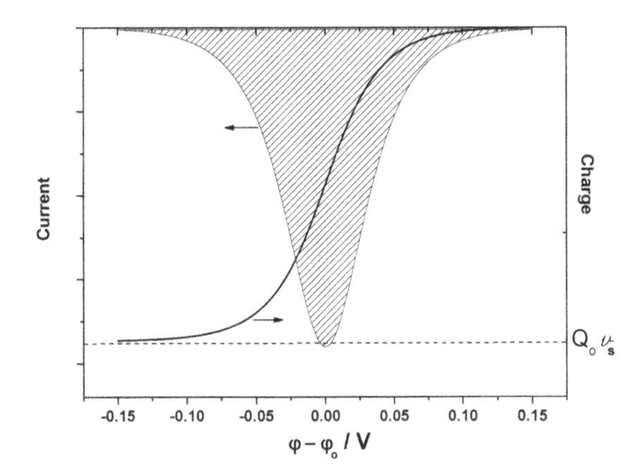

Fig. 6.2. Current and charge versus potential for a Langmuir isotherm whose potential dependence is given by Eq. (6.9). Such curves are obtained by a slow potential sweep. The absolute value of the current depends on the sweep rate (see text).

of the adsorbates is quite important; in addition, the adsorbed ions need not be totally discharged – a point which we will discuss further – and Eq. (6.8) need not hold. The simple adsorption isotherm Eq. (6.9) should rather be viewed as an ideal reference situation.

A simple way to study the potential dependence of an adsorption reaction is a *potential sweep*. In this procedure the electrode potential is first kept in a region where the adsorption is negligible; then the potential is varied slowly and with a constant rate $v_s = d\phi/dt$, and the resulting current i is measured. The sweep rate v_s must be chosen with care. It must be so slow that the reaction is in equilibrium, and that the current due to the charging of the double layer is negligible or small. On the other hand, v_s must be so large that a sizable current flows. Sweep rates of the order of a few mV s^{-1} or somewhat faster are common. In simple cases the current is proportional to the rate of change of the coverage:

$$I = Q_0 \frac{d\theta}{dt} \tag{6.10}$$

where Q_0 is the charge required to form a monolayer of the adsorbate. This equation only holds if the charge required to adsorb one particle is independent of the coverage. This need not be the case; at small coverages the adsorbate may still be charged, while at high coverages Coulomb repulsion prevents the accumulation of a sizable charge on the interface, and the adsorbates will be discharged. Figure 6.2 shows the form of the current-potential curve if both Eqs. (6.9) and (6.10) hold; the absolute value of the current depends on the sweep rate and on Q_0. Again, this curve should be viewed as an ideal reference case, and real curves will differ significantly. A repulsive adsorbate interaction, for example, will broaden the peak in Fig. 6.2, while an attractive interaction will lead to narrow peaks. If the charge per adsorbed particle is constant, the coverage at a given potential can be determined by measuring the charge that has flowed:

$$\theta(\phi) = \frac{Q(\phi)}{Q_0} = \frac{1}{Q_0} \int_{\phi_1}^{\phi} \frac{I}{v_s} d\phi \tag{6.11}$$

where the potential ϕ_1 has to be in the region where the species is not adsorbed.

Other phenomena such as *phase formation* or *phase transitions* will also show up in such current-potential curves. It will be apparent by now that adsorption is a complicated process; only a few systems are well understood. We will consider a few illustrative examples later in this chapter, and defer the more complicated theoretical aspects to later chapters.

6.3 The dipole moment of an adsorbed ion

In general a polar bond is formed when an ion is specifically adsorbed on a metal electrode; this results in an uneven distribution of charges between the adsorbate and the metal and hence in the formation of a surface dipole moment. So the adsorption of an ion gives rise to a dipole potential drop across the interface in addition to that which exists at the bare metal surface.

Electrode	Ion	$\mu \times 10^{-30}$ Cm
Hg	Rb^+	4.07
Ga	Rb^+	0.90
Hg	Cs^+	4.65
Ga	Cs^+	0.90
Hg	Cl^-	−3.84
Hg	Br^-	−3.17
Hg	I^-	−2.64

Table 6.1. Dipole moments of a few ions adsorbed from an aqueous solution at low coverage.

The same effect exists for adsorption on a metal surface from the gas phase. In this case the adsorbate-induced dipole potential changes the work function by an amount $\Delta\Phi$. If n_{ad} is the number of adsorbed molecules per unit area, the component μ_x of the dipole moment of single adsorbed molecule can be inferred from the relation:

$$\Delta\Phi = \frac{n_{ad}\mu_x}{\epsilon_0} \tag{6.12}$$

As before, the x direction has been taken normal to the metal surface. In electrochemistry, the dipole moment μ_x associated with an adsorbate bond can be defined in the following way: For simplicity suppose that the electrode has unit area. At the beginning the electrode surface is bare and kept at the pzc. Then a number $n_{\rm ad}$ of ions with charge number z are adsorbed; simultaneously a counter charge $-ze_0 n_{\rm ad}$ is allowed to flow onto the metal surface. The change $\Delta\phi$ in the electrode potential is related to the dipole moment through:

$$e_0 \, \Delta\phi = \frac{n_{\rm ad}\mu_x}{\epsilon_0} \tag{6.13}$$

Note that both before and after the experiment the sum of the charges on the metal surface and in the adsorbate layer is zero, and hence there is no excess charge in the diffuse part of the double layer. However, after the adsorption has occurred, the electrode surface is no longer at the pzc, since it has taken up charge in the process.

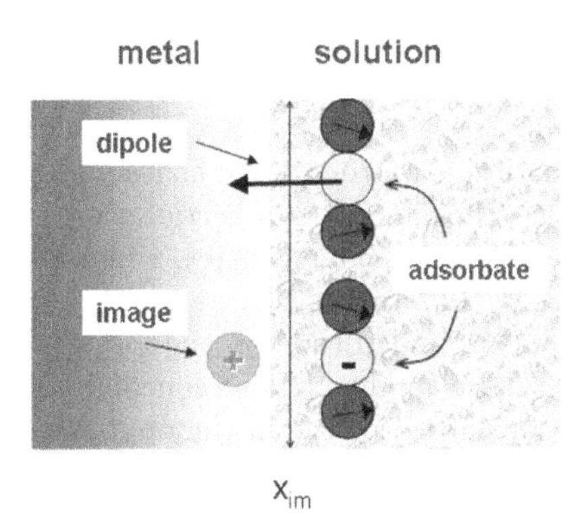

Fig. 6.3. Two alternative ways of viewing the charge distribution in an adsorption bond. The *upper part* of this figure shows the dipole moment; the *lower part* shows a partially charged adsorbate and its image charge. The dipole moments of the surrounding solvent molecules are oriented in the direction opposite to the adsorbate dipole.

In the gas phase the dipole moment determined through Eq. (6.12) refers to an individual adsorbed particle. This is not so in the electrochemical situation. The dipole moment of an adsorbed species will tend to align neighboring solvent molecules in the opposite direction, thereby reducing the total dipole potential drop (see Fig. 6.3). Only the total change in dipole potential can be measured, and there is no way of dividing this into separate contributions from the adsorbate bond and the reorientation of the solvent. A few values of such electrochemical dipole moments are given in Table 6.1. For comparison we note that the dipole moments of alkali ions adsorbed from the vacuum are

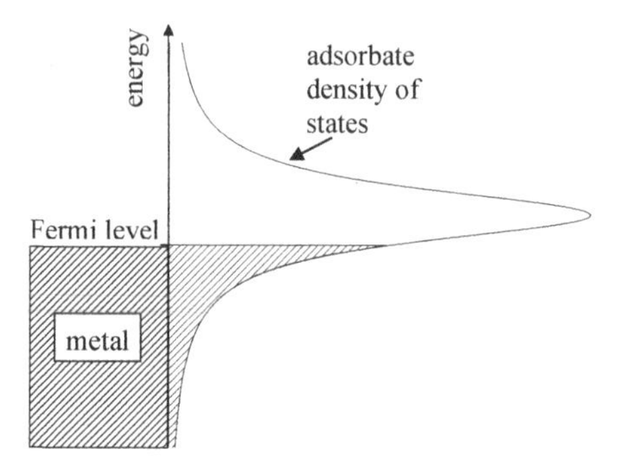

Fig. 6.4. Density of states of an adsorbed cation (schematic).

usually of the order of the order of 10^{-29} Cm. Because of the screening by the solvent, the apparent dipole moment of an ion adsorbed from a solution on a particular metal is often substantially smaller than that of the same ion adsorbed in the vacuum.

If the adsorbate bond has a strong ionic character, as is the case for the alkali and the halide ions, the concept of a *partial charge* is useful. One thinks of the adsorbed ion as carrying a charge $z_{\rm ad}e_0$, which is generally fractional, i.e. $z_{\rm ad}$ is not an integer. The excess charge on the ion induces an image charge of equal and opposite magnitude on the metal surface (see Fig. 6.3), resulting in a surface dipole moment. The concomitant potential reorients the neighboring polar solvent molecules in the opposite direction so that the dipole potential induced by the adsorbate is reduced by the dipole moments of the solvent molecules. A related concept is that of a *partial charge transfer coefficient* l, which is defined as $l = z_{\rm ion} - z_{\rm ad}$.

In principle the partial charge on an adsorbate is ill defined, since one has to introduce a plane separating the electronic density into a part belonging to the adsorbate and one belonging to the metal; obviously, it is not measurable. However, the notion of partial charge can be understood in terms of quantum-mechanical considerations. To be specific, let us consider the adsorption of a Cs^+ ion from an aqueous solution, and assume that the electrode potential is in the range where no reactions occur. When the ion is in the bulk of the solution the valence orbital has a well-defined energy lying above the Fermi level of the electrode; hence the valence orbital is empty. When this ion is adsorbed on the metal electrode, its valence orbital overlaps with the metal orbitals. If we put an electron into the valence orbital of the adsorbed Cs atom, it has only a finite lifetime τ in this state before it is transferred to the metal; the stronger the interaction, the shorter is τ. According to the Heisenberg uncertainty principle, a finite lifetime τ entails an energy uncertainty $\Delta = \hbar/\tau$. Hence the valence orbital is broadened and acquires a *density of states* $\rho(\epsilon)$ of width

Δ, a phenomenon known as *lifetime broadening* and familiar from electronic spectroscopy. This density of states is filled up to the Fermi level of the metal. For an adsorbed Cs atom the center of the density of states lies well above the Fermi level E_F (see Fig. 6.4), the occupancy n is generally quite small, and the partial charge $z_{ad} = 1 - n$ is close to unity. In contrast, halide ions typically carry a negative excess charge, and the center of the density of states of their valence orbitals lies below or near the Fermi level of the metal.

6.4 Electrosorption valence

The distribution of charges on an adsorbate is important in several respects: It indicates the nature of the adsorption bond, whether it is mainly ionic or covalent. Therefore, a fundamental problem of classical electrochemistry is: What does the current associated with an adsorption reaction tell us about the charge distribution in the adsorption bond? Ultimately the answer is a little disappointing: All the quantities that can be measured do not refer to an individual adsorption bond, but involve also the reorientation of solvent molecules and the distribution of the electrostatic potential at the interface. This is not surprising; after all, the current is a macroscopic quantity, which is determined by all rearrangement processes at the interface. An interpretation in terms of microscopic quantities can only be based on a specific model. Therefore, DFT calculations, especially if they include some water besides the electrode and the adsorbate, are especially valuable to understand the adsorption bond.

There is a formal similarity between adsorption and reactions such as metal deposition which gives rise to the concept of *electrosorption valence*. Consider the deposition of a metal ion of charge number z on an electrode of the same material. If the electrode potential ϕ is kept constant, the current density j is:

$$j = -ze_0 \left(\frac{\partial N}{\partial t} \right)_\phi \tag{6.14}$$

where N is the number of particles deposited per unit area. Likewise, when an adsorbate (index i) is deposited on a metal electrode, the resulting current will be proportional to the adsorption rate:

$$j = -le_0 \left(\frac{\partial \Gamma_i}{\partial t} \right)_{\phi, \Gamma_j \neq \Gamma_i} \tag{6.15}$$

where Γ_i is the amount adsorbed, also called the surface excess the surface excess of species i; this includes any excess present in the diffuse layer – a precise definition will be given in Chap. 8. By using $d\sigma = j\ dt$ and rearranging:

$$\left(\frac{\partial \sigma}{\partial \Gamma_i} \right)_{\phi, \Gamma_j \neq \Gamma_i} = le_0 \tag{6.16}$$

where σ is the surface charge density on the metal. The coefficient l was given different names by different authors,[1] but the term *electrosorption valence*, coined by Vetter and Schultze [2], has stuck. Equivalently, it can be defined through:

$$le_0 = \left(\frac{\partial \mu_i^s}{\partial \phi} \right)_{\Gamma_i} \tag{6.17}$$

We defer the proof to Chap. 8, which contains the required thermodynamic relations.

The definition of the electrosorption valence involves the total surface excess, not only the amount that is specifically adsorbed. It is common to correct the surface excess Γ_i for any amount that may be in the diffuse double layer in order to obtain the amount that is specifically adsorbed. This can be done by calculating the excess in the diffuse layer from the Gouy–Chapman theory. Often this correction is small, particularly if the species is strongly adsorbed, or if an excess of a nonadsorbing electrolyte is used. However, if the correction term is large, Eq. (6.17) need not hold for the corrected quantity, since this equality has been proved only for the total excess.

The interpretation of the electrosorption valence is difficult. The following, somewhat naive argument shows that it involves both the distribution of the potential and the amount of charge transferred during the adsorption process. Suppose that an ion S^z is adsorbed and takes up λ electrons in the process. λ need not be an integer since there can be partial charge transfer. We can then write the adsorption reaction formally as:

$$S^z \rightarrow S^{z+\lambda} + \lambda e^- \tag{6.18}$$

As noted before, the partial charge transfer is not well defined. Nevertheless, let us suppose that we can treat $S^{z+\lambda}$ like a normal species. Its electrochemical potential is then:

$$\tilde{\mu}_i^{\text{ad}} = \mu_i^{\text{ad}} + (z + \lambda) e_0 \phi_{\text{ad}} \tag{6.19}$$

where ϕ_{ad} denotes the electrostatic potential at the adsorption site. Since the reaction is in equilibrium, the electrochemical potentials must balance. Setting the electrostatic potential in the solution equal to zero, we obtain:

$$\mu_i^s = \mu_i^{\text{ad}} + (z + \lambda) e_0 \phi_{\text{ad}} - \lambda e_0 \phi_m \tag{6.20}$$

where ϕ_m is the potential of the metal. Differentiating with respect to the electrode potential ϕ, which differs from ϕ_m by a constant, gives:

$$l = gz - \lambda(1 - g), \qquad \text{where } g = \left(\frac{\partial \phi_{\text{ad}}}{\partial \phi} \right)_{\Gamma_i} \tag{6.21}$$

While this equation is certainly not exact, it can be used for qualitative interpretations. In particular, the following limiting cases are of interest:

---

[1] Usually the electrosorption valence is denoted by γ, which we use for the surface tension. The symbol l was used earlier by Lorenz and Salie [1].

Electrode	Ion	l
Hg	Rb^+	0.15
Ga	Rb^+	0.20
Hg	Cs^+	0.18
Ga	Cs^+	0.20
Hg	Cl^-	−0.20
Hg	Br^-	−0.34
Hg	I^-	−0.45

Table 6.2. Electrosorption valences of a few simple ions at the pzc and at low coverage.

1. Total discharge: $\lambda = -z$, or $l = z$
2. Incorporation into the electrode: $\phi_{\text{ad}} = \phi_m$, $g = 1$, and $l = z$.
3. No charge transfer: $\lambda = 0$, $l = gz$.

In general, the electrosorption valence depends both on the electrode potential and on the amount adsorbed, as may be expected from Eq. (6.21). Table 6.2 lists the electrosorption valences of a few simple ions at the pzc and at low coverages on liquid metals [3], where measurements are easier than on solids. The low values for the alkali ions Rb^+ and Cs^+ are generally thought to indicate the absence of partial charge transfer. In contrast, the values for the halide ions may indicate a partial transfer of an electron to the metal. In underpotential deposition of a monolayer of metal ions (see next chapter) the electrosorption valence is generally equal to the charge number of the metal ion, indicating total discharge. A final word of caution: In mixed solutions coadsorption may take place and make a proper determination of the electrosorption valence difficult.

6.5 Electrosorption valence and the dipole moment

The electrosorption valence can be related to the dipole moment of an adsorbed species introduced above. For this purpose consider an electrode surface that is initially at the pzc and free of adsorbate. When a small excess charge density σ is placed on the metal, its potential changes by an amount $\Delta\phi$ given by:

$$\Delta\phi = \frac{\sigma}{C} = \frac{\sigma}{C_{\text{H}}} + \frac{\sigma}{C_{\text{GC}}} \tag{6.22}$$

where we have split the interfacial capacity C into the Gouy–Chapman part C_{GC} and the Helmholtz part C_{H}. Equation (6.22) is a linear expansion in terms of σ. When a small number N_i of a species with charge number z is adsorbed per unit area at fixed σ, the resulting change in the electrode potential is proportional to N_i. The total charge density at the interface is now $\sigma + ze_0N_i \equiv \sigma + \sigma_i$, and this is balanced by the charge in the diffuse layer. So we have:

$$\Delta\phi = B\sigma_i + \frac{\sigma}{C_{\mathrm{H}}} + \frac{\sigma + \sigma_i}{C_{\mathrm{GC}}} \tag{6.23}$$

with an unknown coefficient B. We have made no assumption about the actual distribution of the charge at the interface. The electrosorption valence is:

$$\tilde{l} = -z\left(\frac{\sigma}{\sigma_i}\right)_{\Delta\phi=0} = z\frac{B + 1/C_{\mathrm{GC}}}{1/C_{\mathrm{GC}} + 1/C_{\mathrm{H}}} \tag{6.24}$$

The tilde indicates that the value is not corrected for the diffuse layer. The corrected value is obtained by eliminating the C_{GC} terms:

$$l = zBC_{\mathrm{H}} \tag{6.25}$$

To obtain the dipole moment we set $\sigma = -\sigma_i$ in Eq. (6.23) so that the diffuse double layer is free of excess charge (see Sect. 4.3).

$$\Delta\phi_{\sigma=-\sigma_i} = \sigma_i\left(B - \frac{1}{C_{\mathrm{H}}}\right) \tag{6.26}$$

The dipole moment μ_i per adsorbate is obtained by dividing this potential drop by N_i/ϵ_0, and changing the sign, so that a positive charge on the adsorbate corresponds to a positive dipole moment:

$$\mu_i = -ze_0\epsilon_0\left(B - \frac{1}{C_{\mathrm{H}}}\right) = \frac{ze_0\epsilon_0}{C_{\mathrm{H}}}(1 - l/z) \tag{6.27}$$

which is the desired relation. For a different derivation see [4]. We think that the dipole moment is a more useful quantity than the electrosorption valence since it can be interpreted without recourse to the badly defined concept of partial charge transfer. Even so, μ_i is not the dipole moment of an individual adsorbed molecule. The solvent molecules in the vicinity of the adsorbate will be oriented by the dipole moment of the adsorbate, and the resulting change in the interfacial potential is reflected in μ_i [5].

6.6 Structures of commensurate overlayers on single crystal surfaces

The overlayer structure that result when a species is adsorbed on a single crystalline surface depends on various factors, such as the relative sizes of the adsorbate and substrate atoms (or molecules), and the interactions between the particles involved. If the interactions between the adsorbed species are repulsive, the resulting overlayer often shows a homogeneous structure. On the contrary, if attractive forces exist, there is a tendency to form islands or patches on the surface. Moreover, as we have already discussed in Chap. 2 there are also preferential sites for the adsorption of different species (atop,

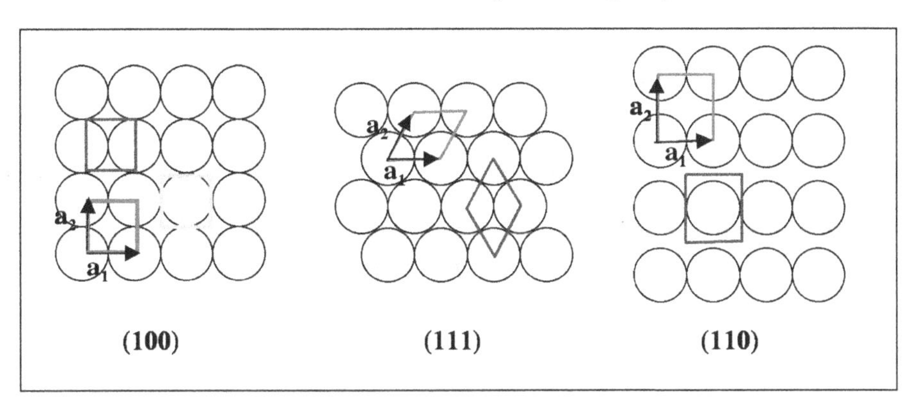

Fig. 6.5. Unit cells of the principle planes of an fcc crystal.

bridge, threefold sites, etc.). In this section we shall look at the formation of commensurate overlayers, which we shall define below, and the notation to represent them.

Single crystals have a periodically ordered arrangement of atoms at the surface. The structure depends on the direction of cutting the crystal as already mentioned in Chap. 2. Just like the structure of three-dimensional crystals is defined by three-dimensional unit cells, the structure of a single-crystal surface can be characterized by a two-dimensional unit cell, which is defined as the simplest periodically repeating unit which can be identified in an ordered two-dimensional array. Thus, the whole surface structure can be constructed by repeated translation of the unit cell. As examples we show in Fig. 6.5 the three principal lattices planes of a fcc crystal and several possible choices for the unit cell are indicated. Of course, all legitimate choices of the unit cell give rise to the same lattice.

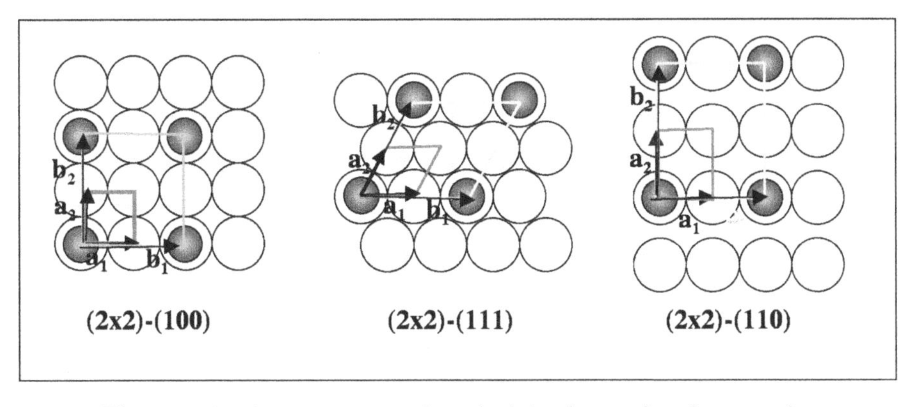

Fig. 6.6. 2×2 structures on the principle planes of an fcc crystal.

Two vectors $\mathbf{a}_1$ and $\mathbf{a}_2$ with a common origin are usually selected to define the unit cell. In the case of ordered overlayers formed by adsorbates we can employ the same procedure as before and define a unit cell for the overlayer through vectors $\mathbf{b}_1$ and $\mathbf{b}_2$. When the substrate and the adsorbate structures are related by a simple mathematical transformation, one speaks of a *commensurate* structure. otherwise it is *incommensurate.* Frequently, Wood's notation is employed when the vectors form the same angle for the two unit cells that corresponding to the substrate and to the overlayer. The lengths of the two vectors $\mathbf{b}_1$ and $\mathbf{b}_2$ are expressed in terms of $\mathbf{a}_1$ and $\mathbf{a}_2$, respectively, as $(|\mathbf{b}_1/|\mathbf{a}_1| \times |\mathbf{b}_2/|\mathbf{a}_2|)$. As an example, Fig. 6.6 shows the (2×2) structures on the three principal lattice planes of a fcc crystal. Here the adsorbate is smaller than the substrate atom and adsorbed on an atop site, but the same concepts can be applied for other sites and sizes. Often, the letter p precedes the description of the structure $p(2 \times 2)$ to indicate that it is a primitive structure, the simplest possible unit, and to distinguish it from the closely related $c(2 \times 2)$ structure. The latter actually does not correspond to a primitive cell because it has an additional species in the centre of the (2×2) structure. The corresponding primitive cell for this structure is better described as $(\sqrt{2} \times \sqrt{2})R45$, which means that the vectors b_1 and b_2 are $\sqrt{2}$ larger than $\sqrt{a}_1$ and $\sqrt{a}_2$, respectively, and that the unit cell of the overlayer is rotated 45 degrees with respect to the substrate unit cell (see Fig. 6.7). Another typical overlayer structure which is frequently observed on the (111) surface, also shown in Fig. 6.7, is that represented by the primitive cell $(\sqrt{3} \times \sqrt{3})R30$.

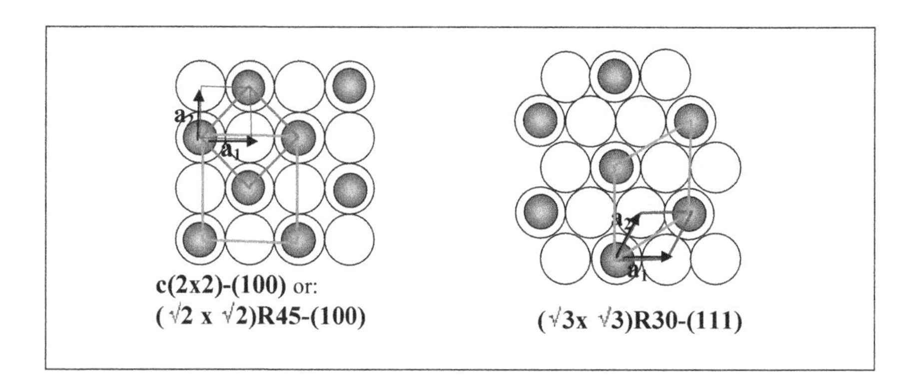

Fig. 6.7. Two examples of rotated structures.

References

1. W. Lorenz and G. Salie, Z. *Phys. Chem. NF* **29** (1961) 390, 408.
2. K.J. Vetter and J.W. Schultze, *Ber. Bunsenges. Phys. Chem.* **76** (1972) 920, 927.

3. J.W. Schultze and F.D. Koppitz, *Electrochim. Acta* **21** (1977) 81.
4. K. Bange, B. Strachler, J.K. Sass, and R. Parsons, *J. Electroanal. Chem.* **229** (1987) 87.
5. W. Schmickler and R. Guidelli, *J. Electroanal. Chem.* **235** (1987) 387; W. Schmickler, ibid. **249** (1988) 25.

7

Adsorption on metal electrodes: examples

7.1 The adsorption of halides on metal electrodes

The adsorption of halides on single crystal metals is a paradigmatic example of the competition between the interactions adsorbate – substrate and adsorbate – adsorbate. Because of their weak solvation shells, anions adsorb easily on metal surfaces, particularly at potentials positive of the pzc. However, it is not easy to determine the coverage. Often the adsorbed ions carry a partial charge, they repel each other, and the coverage increases only slowly with increasing electrode potential. This makes it difficult to determine the coverage by measuring the charge that flows during a potential sweep or an adsorption transient. In systems where the adsorbed anions form a regular lattice, the structure can be elucidated by local probe techniques such as the scanning tunneling microscope (STM), or X-ray scattering and spectroscopy using synchrotron radiation, which allow direct studies of the electrochemical interface on the atomic and nanometer scale.

We briefly analyze here as examples the adsorption of chloride and bromide on Ag(100). In the case of chloride adsorption, the sharp peak at -0.5 V vs. SCE observed in the voltammograms and also in the capacity curves can be linked to the transition from a disordered to an ordered c($2\times$ 2) phase (see Fig. 7.1), which has been observed by X-ray. For bromide adsorption, the same transition can been seen at a lower potential (-0.75 V). Because of its weaker solvation shell bromide is more strongly adsorbed than chloride, and hence, the transition occurs at a lower charge density.

Another case in point is the adsorption of iodide on a Pt(111) electrode. Platinum forms a face-centered cubic lattice with a lattice constant of $a = 3.92$ Å. The (111) surface has a triangular lattice structure with a nearest-neighbor distance of $a/\sqrt{2} = 2.77$ Å. When this surface is immersed in an aqueous solution containing iodide, the latter is adsorbed over a wide range of potentials. Topographic images of a regular adsorbate lattice were obtained with a scanning tunneling microscope under the following conditions: a concentration of 10^{-4} M KI, and 10^{-2} $NaClO_4$ plus 10^{-4} M $HClO_4$ as supporting electrolytes,

W. Schmickler, E. Santos, *Interfacial Electrochemistry*, 2nd ed.,
DOI 10.1007/978-3-642-04937-8_7, © Springer-Verlag Berlin Heidelberg 2010

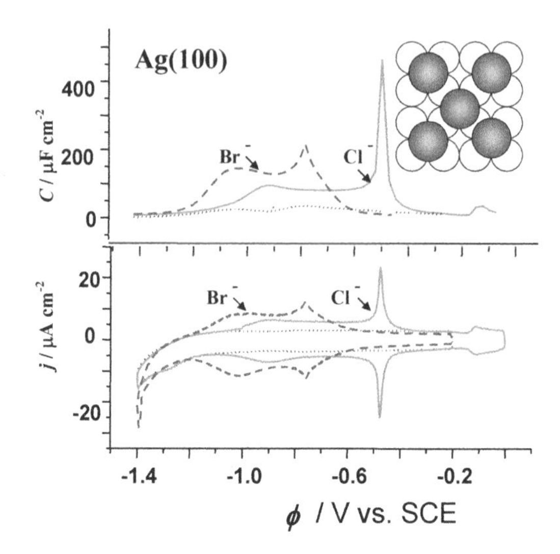

Fig. 7.1. Capacity curves (*above*) and cyclic voltammogram for the adsorption of chloride and bromide on Ag(100). The *dotted curve* is for the ClO_4^- anion, which is only weakly adsorbed. Data taken from [1].

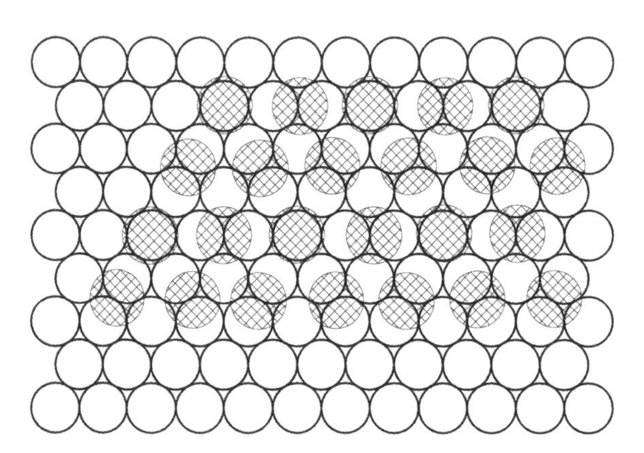

Fig. 7.2. Adsorption of iodide (*hatched circles*) on Pt(111) (*open circles*). For greater clarity only part of the adsorbate lattice is shown.

and an electrode potential of 0.9 V vs. RHE [2] (RHE stands for *reversible hydrogen electrode*; so the potential is referred to the equilibrium potential of the hydrogen evolution reaction in the same solution). The observed structure is shown in Fig. 7.2. The iodide lattice is also hexagonal, with a nearest-neighbor distance of 4.16 Å. Further examination showed that the adsorbed ions are not all equivalent: 1/4 of the ions are positioned a little higher, each such ion being surrounded by a hexagon of ions sitting a little lower. This is compatible with the structure shown in the figure, in which 1/4 of the ions sit at atop sites, and are surrounded by ions sitting at bridge sites. The total coverage is 4/9 of the coverage calculated for the case in which one iodide would be adsorbed on each platinum atom. However, such a one-to-one correspondence between iodide and platinum is not attainable in practice since the I^- ion, which has a radius of 2.16 Å, is larger than a platinum atom. This structure indicates that the interaction between the adsorbed particles is repulsive, probably because they carry a small negative charge. In contrast a strong attractive interaction leads to the formation of islands of adatoms at low coverages.

7.2 Underpotential deposition

The adsorption of metal ions on a foreign metal substrate is a particularly intriguing topic. There are two possibilities: The Gibbs energy of interaction of the adsorbate with the substrate can be weaker or stronger than the adsorbate-adsorbate interaction. In the first case the adsorbate will be deposited at potentials lower than the equilibrium potential ϕ_{00} for bulk deposition and dissolution, and will often form three-dimensional clusters from the start. In the latter case the adsorbate can be deposited on the foreign substrate at potentials *above* ϕ_{00}. This case is known as *underpotential deposition* (upd), a prime example for confusing terminology. Generally up to a monolayer can be formed in this way; in a few cases the adsorbate-substrate interaction is sufficiently strong to allow the deposition of a second layer at potentials slightly above ϕ_{00}.

The energetic aspects of underpotential deposition can be investigated by a slow (i.e., a few millivolts per second) potential scan starting at a potential so high that no adsorption takes place. As the potential is lowered, one or more current peaks are observed, which are caused by the adsorption of the metal ions (see Fig. 7.3). According to the usual convention, the adsorption current is negative (i.e., cathodic). Different peaks may correspond to different adsorption sites, or to different structures of the adsorbate layer. If the potential is scanned further past the equilibrium potential ϕ_{00}, the usual bulk deposition is observed.

Instead of performing a single potential scan, it is common to reverse the direction of the scan at the beginning of the bulk deposition, and to observe the desorption of the adatoms, which gives rise to positive current peaks. The sweep direction is then reversed again at a potential well positive of the desorption, and this procedure is repeated several times. The resulting current-potential curve is called a *cyclic voltammogram*; further details of this technique will be given in Chap. 19. Successive sweeps give identical curves if the reactions that take place in this range are reversible. If the sweep rate is slow and the adsorption reaction reversible, the adsorption and desorption peaks are at the same potential; this is almost, but not quite, the case for the

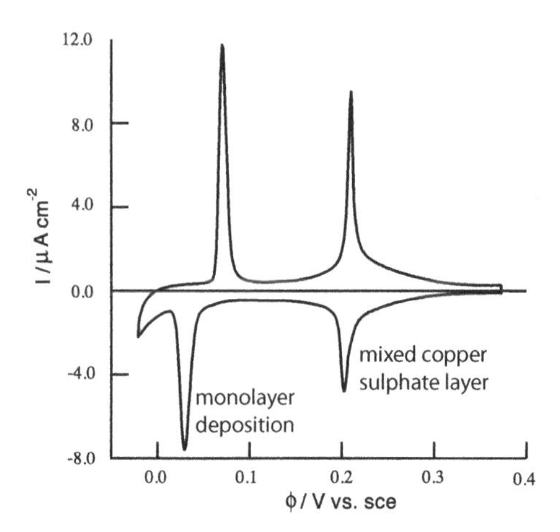

Fig. 7.3. Cyclic voltammogram for the upd of copper on Au(111); the electrolyte is an aqueous solution of 0.05 M H_2SO_4 and 10^{-3} M $CuSO_4$. Scan rate: 1 mV/s; data taken from [3].

peaks near 0.2 V in the cyclic voltammogram for the upd of Cu^{2+} on Au(111) shown in Fig. 7.3. The charge under the first peak corresponds to the formation of mixed layer of copper and sulphate. We will discuss its structure in detail below. Such a coadsorption of metal ions and anions is quite common in upd. Only at the second peak near 0.03 V a monolayer of copper is adsorbed. Note that the corresponding desorption peak is shifted toward a higher potential (near 0.07 V), possibly because the desorption is very slow.

The difference between the potential of the current peak for the desorption and the bulk deposition potential is known as the *underpotential shift* ϕ_{upd}. For simple systems the value of ϕ_{upd} is independent of the concentration of ions in the bulk of the solution, since the Gibbs energies of adsorption and deposition shift both according to the Nernst equation. Deviations from this behavior may indicate coadsorption of other ions.

During the 1960s and 1970s, before the preparation of single crystals was well established in electrochemistry, many cases of upd were investigated on polycrystalline metals. Surprisingly large upd shifts, up to 1 V in aqueous and even higher in non-aqueous solutions, were observed, which correlated quite well with the difference in the work functions of the two metals involved [3]. No such correlation holds for single crystals, and the data in non-aqueous solvents were found to be a misinterpretation. Further investigations with STM and with x-ray techniques revealed, that several systems that had been thought to be pure metal adsorbate layers really really consisted of co-adsorbed metal and anions, structures akin to two-dimensional salts. Therefore, the correlation between the upd shift and the work function on polycrystalline metals is dubious, and we shall not discuss it further.

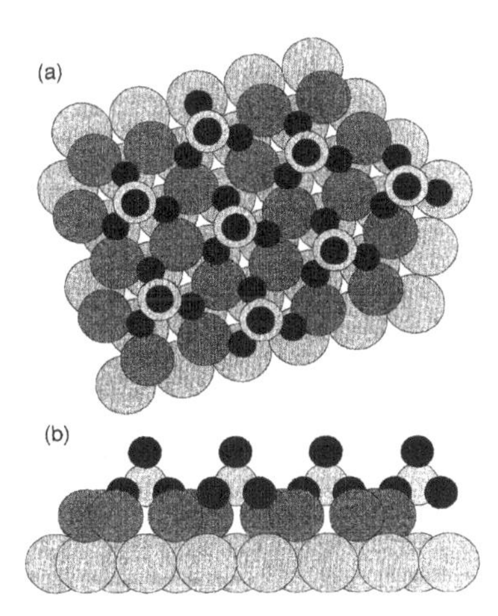

Fig. 7.4. Coadsorbed layer of copper and sulphate on Au(111) (schematic); (**a**) top view, (**b**) side view. The large *light grey* spheres at the *bottom* represent the gold atoms; the copper atoms shown as *medium grey*, the sulfur atoms as small *light grey*, and the oxygen atoms as *dark spheres*. Taken from Toney et al. [4], courtesy of Physical Reviews.

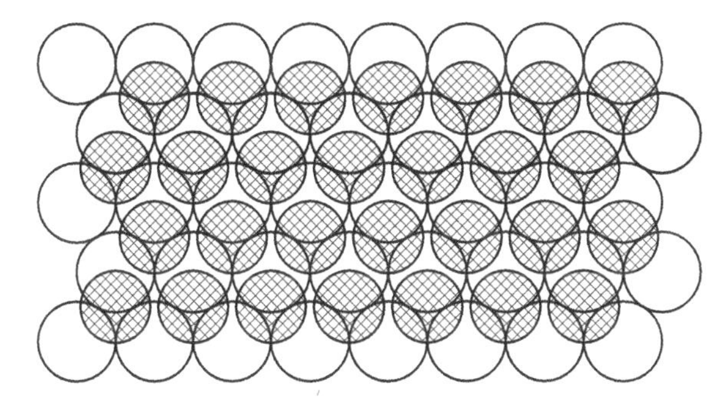

Fig. 7.5. Adsorbate structure of a monolayer of copper on Au(111).

To obtain structural information on the adsorption sites, single crystal electrodes must be used. As an example we consider again the upd of copper on Au(111). The adsorption and desorption peaks in Fig. 7.3 are very narrow, as is often the case when regular adsorbate lattices are formed. We look at the deposition process, which corresponds to the negative current in the voltamogram. At potentials below 0.2 V, the surface is covered by a coadsorption layer consisting of 2/3 of a monolayer of Cu and 1/3 of a monolayer of SO_4^{2-} or HSO_4^- ions. The structure, which is technically known as a $\sqrt{3} \times \sqrt{3}R30°$, is shown in Fig. 7.4.

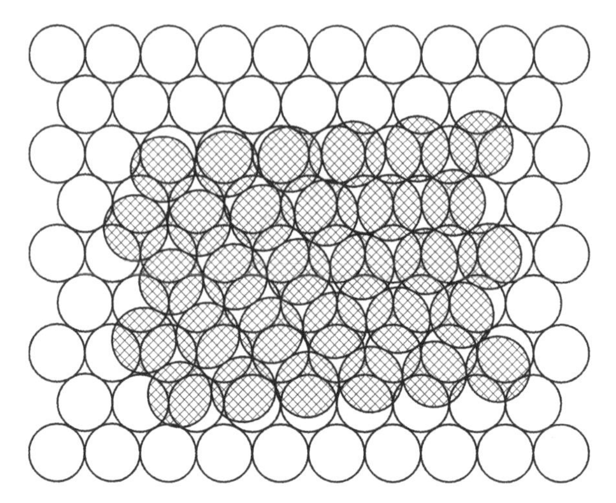

Fig. 7.6. Adsorption of a monolayer of lead (*hatched circles*) on Ag(111) (*open circles*).

At potentials below 0.03 V, the gold surface is covered by a monolayer of copper, whose structure has been elucidated by x-ray absorption spectroscopy [5]. Gold has an fcc lattice, and the Au(111) surface forms a triangular lattice with a lattice constant of 2.89 Å. Copper atoms are smaller than gold atoms, and they adsorb in the threefold hollow sites (see Fig. 7.5), forming a triangular lattice *commensurate* with that of the substrate; i.e., the lattices of the adsorbate and of the surface layer of the substrate are related by a simple mathematical transformation – otherwise the adsorbate lattice is said to be *incommensurate.*

Silver forms an fcc lattice, too, and its lattice constant is almost the same as that of gold. When a Ag(111) surface is immersed in a solution containing a small concentration of Pb^{2+} ions and an inert electrolyte, a potential scan shows a series of upd peaks at potentials near -0.34 V vs. sce (saturated calomel electrode). X-ray scattering [4] showed that in the region negative to these peaks a dense, incommensurate layer of Pb(111) exists whose lattice constant is larger than that of the silver substrate, and whose axis is rotated by 4.5° (see Fig. 7.6).

The two examples discussed here are typical in the sense that metal adsorbates with atoms that are smaller than those of the substrate tend to form commensurate layers, while adsorbates with bigger atoms tend to form incommensurate monolayers [6]. Also, pure upd layers tend to form close to the bulk deposition potential, while structures at higher potentials are usually mixed layers containing both the metal ions and anions.

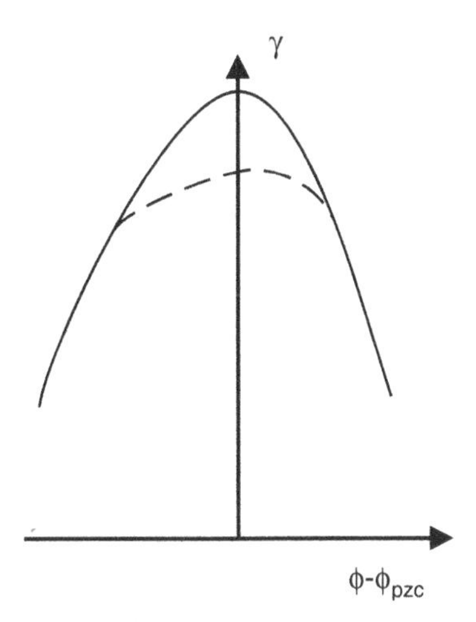

Fig. 7.7. Surface tension of mercury in the presence (*dashed line*) and in the absence (*solid line*) of an aliphatic compound (schematic).

7.3 Adsorption of aliphatic molecules

The adsorption of organic molecules offers a rich phenomenology. A large number of studies have been performed on mercury electrodes, where the surface tension can be measured directly, and the surface charge and the capacity obtained by differentiation. We will not attempt to survey the literature, but consider a simple example: the adsorption of aliphatic compounds.

When the surface tension of a mercury electrode in contact with an aqueous solution containing a neutral aliphatic compound is measured as a function of the electrode charge or potential, the following behavior is observed: The surface tension is substantially lowered in a region embracing the pzc of the electrode in the presence of the adsorbate, while at potentials far from the pzc the surface tension is unchanged (see Fig. 7.7). Obviously, the adsorption of the compound is limited to a region near the pzc. A possible explanation is this: On the one hand, the aliphatic chains are squeezed out toward the surface by the hydrogen-bonded water structure; hence they are adsorbed near the pzc. On the other hand, the dipole moment of an aliphatic compound is lower than that of water; when the charge on the electrode surface is high, the polar water molecules are drawn toward the surface by electrostatic forces, and expel the adsorbed molecules.

The *parallel-capacitor model* suggested by Frumkin [7] is an attempt to turn this into a quantitative argument; we discuss a simplified version. According to this model, the surface consists of patches covered by the adsorbate and patches which are free. Since the dipole moment of water is higher than that of the adsorbate, and the water molecule is smaller, these patches will have a different interfacial capacity per unit area. The interface behaves like two capacitors in parallel. We consider an electrode with unit area, and denote the capacity per unit area of the free surface by C_0, and that of the adsorbate-covered surface by C_1, with $C_1 < C_0$. To simplify the mathematics we assume that C_0 and C_1 are constant, and that the pzc ϕ_{pzc} is not shifted by the presence of the adsorbate. The total charge on the electrode surface is then:

$$\sigma = (1-\theta)C_0(\phi - \phi_{\text{pzc}}) + \theta C_1(\phi - \phi_{\text{pzc}}) \tag{7.1}$$

where θ is the coverage of the adsorbate.

In the Frumkin isotherm Eq. (6.6) we want to single out the contribution of the electrode potential, and write:

$$\frac{\theta}{1-\theta} = \frac{c}{c_0} \exp\left(\frac{\mu_{\text{sol}} - \mu_{\text{ad}}}{kT}\right) \mathrm{e}^{-g\theta} = \frac{c}{c_0} B_0 \exp\left(-\frac{W(\phi)}{kT}\right) \mathrm{e}^{-g\theta} \tag{7.2}$$

where B_0 is independent of the electrode potential and $W(\phi)$ is the electrostatic work required to adsorb a single molecule at a given potential on the electrode surface. When a molecule is adsorbed, the interfacial capacity changes; so we need the work required to change the capacity of a condenser at constant potential ϕ. From simple electrostatics the energy stored in a condenser is $\phi\sigma/2 = C\phi^2/2$, where σ is the charge on one plate. When the capacity is changed by dC, this energy changes by an amount:

$$dW_1 = \frac{1}{2}\phi^2 \, dC \tag{7.3}$$

In addition, the charge on the capacitor changes by an amount $d\sigma = \phi \, dC$, and the potentiostat has to perform work $dW_2 = \phi d\sigma = \phi^2 \, dC$ on the capacitor; so the total change in the energy of the capacitor is:

$$dW = dW_1 - dW_2 = -\frac{1}{2}\phi^2 \, dC \tag{7.4}$$

A moment's thought shows that the minus sign is correct: Increasing the capacity of a condenser at constant potential, for example, by decreasing the plate and hence the charge separation, must lower the energy of the system.

When a single molecule is adsorbed on the surface, the coverage changes by an infinitesimal amount $\Delta\theta = 1/N_{\text{max}}$, where N_{max} is the maximum number of particles that can be adsorbed. From Eq. (7.1) the concomitant change in the capacity is:

$$\Delta C = (C_1 - C_0) \, \Delta\theta = \frac{(C_1 - C_0)}{N_{\text{max}}} \tag{7.5}$$

Substituting this into Eq. (7.4) gives for the electrostatic work required to adsorb one particle:

$$W(\phi) = -\frac{1}{2}(\phi - \phi_{\text{pzc}})^2 \frac{(C_1 - C_0)}{N_{\text{max}}} \tag{7.6}$$

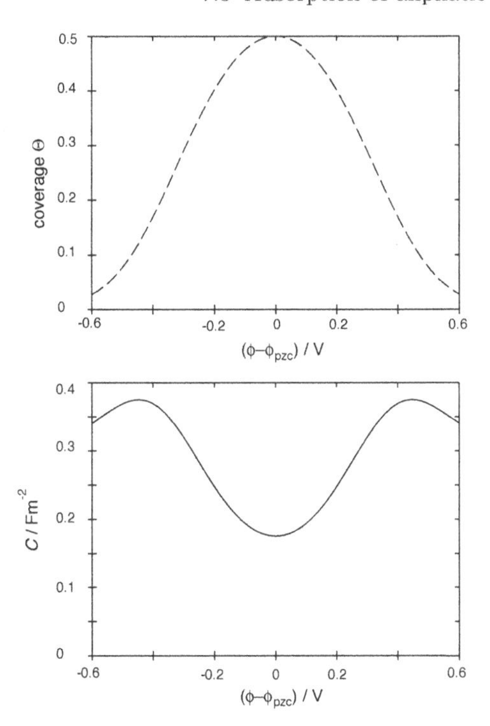

Fig. 7.8. Coverage (**a**) and differential capacity (**b**) for the adsorption of an aliphatic compound according to Eq. (7.7)

Since $C_1 < C_0$, this work is positive, and the coverage decreases away from the pzc. Equations (7.2) and (7.6) can be combined with the Frumkin isotherm, resulting in:

$$\frac{\theta}{1-\theta} = \frac{c}{c_0} B_0 \exp\left(-\frac{(\phi - \phi_{\text{pzc}})^2 (C_1 - C_0)}{2N_{\text{max}}kT}\right) \exp(-\gamma\theta) \qquad (7.7)$$

Typical plots for the coverage and the capacity as a function of the electrode potential are shown in Fig. 7.8. Note the pronounced maxima in the capacity near the potentials where the substance is desorbed. Equation (7.7) can be improved by allowing for the potential dependence of the two capacities C_0 and C_1, and for a shift in the pzc with adsorption, but little is gained in physical insight.

Problems for Chaps. 6 and 7

1. The conventional unit cell of a body-centered cubic crystal (bcc) consists of the eight corners of a cube and the point in the center. Describe the structures of the (100), (111), and (110) planes.

2. A cesium ion has a radius of $r = 1.68$ Å. Assume that it carries a unit charge when it is adsorbed on a plane metal surface, and calculate the dipole moment formed by the ion and its image charge. Calculate the energy of interaction of two Cs^+ ions: (a) when they are touching, (b) when they are 10 Å apart. (c) Assume that the adsorbed Cs^+ ions form a square lattice of lattice constant L. Show that the coverage is given by $\theta = (2r)^2/L^2$. Derive the adsorption isotherm assuming that each ion interacts only with its four nearest neighbors. Assume that the dielectric constant of the surrounding medium is unity.
3. Consider two different metals in contact and assume that both are well described by the Thomas–Fermi model with a decay length of $L_{TF} = 0.5$ Å. (a) Calculate the dipole potential drop at the contact if both metals carry equal and opposite charges of 0.1 Cm^{-2}. (b) If the work functions of the two metals differ by 0.5 eV, how large is the surface-charge density on each metal?
4. Prove that the Frumkin isotherm exhibits a phase transition for $g < -4$. For this purpose, calculate $dc_A/d\theta$ from Eq. (6.6) and determine, under which condition it has a zero.

References

1. G. Beltramo, Dissertation, Universidad Nacional de Cordoba, Argentina, 2001.
2. R. Vogel, I. Kamphausen, and H. Baltruschat, *Ber. Bunsenges. Phys. Chem.* **96** (1992) 525; B.C. Schardt, S.L. Yau, and F. Rinaldi, *Science* **243** (1989) 981.
3. D. Kolb, *Advances of Electrochemistry and Electrochemical Engineering*, Vol. 11, p. 125. Wiley, New York, NY, 1978.
4. M.F. Toney and O. Melroy, *Electrochemical Interfaces*, edited by H.D. Abruña, p. 57. VCH, New York, NY, 1991.
5. H.D. Abruña, *Electrochemical Interfaces*, edited by H.D. Abruña, p. 1. VCH, New York, NY, 1991.
6. J.W. Schultze and D. Dickertmann, *Surf. Sci.* **54** (1976) 489.
7. A.N. Frumkin, *Z. Phys.* **35** (1926) 792.

8

Thermodynamics of ideal polarizable interfaces

8.1 Liquid electrodes

Here we extend on the brief introduction of the surface tension given in Chap. 4. To a large extent, our treatment is based on the works of Grahame [1] and Parsons [2] and, like all proper thermodynamics, is exact. However, our way of defining the surface tension and deriving its differential is different, and is based on a discussion with H. Ibach, whose book contains an alternative derivation directed at solid electrodes [3]. In the classical derivation, the surface tension is introduced ad hoc, and it is not clear that it has the potential as its natural variable. For liquid electrodes thermodynamics offers a precise way to determine the surface charge and the surface excesses of a species. This is one of the reasons why much of the early work in electrochemistry was performed on liquid electrodes, particularly on mercury – another reason is that it is easier to generate clean liquid surfaces than clean solid surfaces. With some caveats and modifications, thermodynamic relations can also be applied to solid surfaces, and it is still the most exact way to obtain surface excesses.

Thermodynamics no longer plays the important role like 50 years ago. Therefore we do not treat it in one of the first chapters, like older textbooks invariably do. Nevertheless, an understanding of thermodynamics is essential; in fact, since the publication of the first edition of this book a fair number of papers have appeared, whose arguments were simply based on false thermodynamics, confusing surface tension with surface free energy or with surface stress. We will first consider the interface between a liquid electrode and an electrolyte solution, and turn to solid electrodes later.

The simplest way to treat an interface is to consider it as a phase with a very small but finite thickness in contact with two homogeneous phases (see Fig. 8.1). The thickness must be so large that it comprises the region where the concentrations of the species differ from their bulk values. It turns out that it does not matter, if a somewhat larger thickness is chosen. For simplicity we assume that the surfaces of the interface are flat. This interface contains

W. Schmickler, E. Santos, *Interfacial Electrochemistry*, 2nd ed.,
DOI 10.1007/978-3-642-04937-8_8, © Springer-Verlag Berlin Heidelberg 2010

a variety of both charged and uncharged particles labelled i. We consider a system at constant temperature and pressure and therefore start from the Gibb's free energy, whose differential for the interface is in standard notation:

$$dG^\sigma = -S^\sigma dT + V^\sigma dp + \sum_i \tilde{\mu}_i^\sigma \, dN_i^\sigma \tag{8.1}$$

where the index σ indicates that the corresponding quantity pertains to the interface. The whole system including the two adjoining bulk phases is supposed to be in thermal and mechanical equilibrium, so that temperature and pressure are constant. To avoid cluttering the equations with indices, we will use the index σ only if it is not obvious that the quantity refers to the interface.

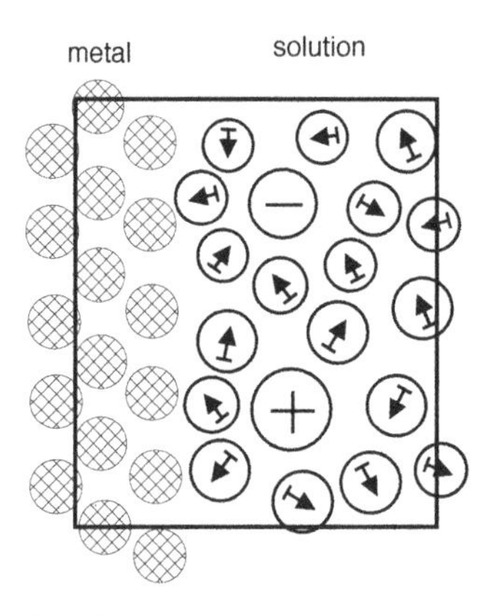

Fig. 8.1. The interface between a metal and an electrolyte solution.

Equation (8.1) does not contain the surface charge or the potential explicitly. The charge is hidden in the particle numbers of the ions, the potential in the electrochemical potentials $\tilde{\mu}$. Since G has the particle number as variable, it implicitly depends on the charge. In accord with our considerations of Chap. 4 we have to perform a transformation to obtain a thermodynamic potential which has the potential, or the $\tilde{\mu}$, as variables, since in electrochemistry the interface is kept at constant potential. Therefore we define:

$$X^\sigma = G^\sigma - \sum_i \tilde{\mu}_i^\sigma \, N_i^\sigma \tag{8.2}$$

Differentiation gives:

$$dX^\sigma = -S^\sigma \, dT + V^\sigma \, dp - \sum_i N_i^\sigma \, d\tilde{\mu}_i^\sigma \tag{8.3}$$

The surface tension is defined as $\gamma = X^\sigma / A$, where A is the surface area. We specialize to the case of constant temperature, pressure, and surface area, and introduce the surface concentrations:

$$\Gamma_i^* = \frac{N_i^\sigma}{A^\sigma} \tag{8.4}$$

and obtain the *Gibbs adsorption equation*:

$$d\gamma = -\sum_i \Gamma_i^* \, d\tilde{\mu}_i^\sigma \tag{8.5}$$

The interface is in contact with two bulk phases, the metal electrode (index m) and the solution (index s).[1] Formally, we consider the metal to be composed of metal atoms M, metal ions M^{z+}, and electrons e^-; these particles are present both in the electrode and the interface, but not in the solution. On the other hand, certain cations and anions and neutral species occur both in the solution and the interface. Since the electrode is ideally polarizable, no charged species can pass through the interface.

The surface concentrations Γ_i^* depend on the thickness of the interfacial region, and we would like to express them through quantities which are independent of it. This can be done for those species which occur both at the interface and in the solution. Usually one of the components of the solution, the solvent, has a much higher concentration then the others. We denote it by the index "0", and introduce *surface excesses* with respect to the solvent in the following way: In the bulk of the solution the Gibbs–Duhem equation (at constant T and p) is simply $\sum N_i \, d\tilde{\mu}_i = 0$, or:

$$d\tilde{\mu}_0^s = -\sum_i^{sol}{}' \frac{N_i^s}{N_0^s} \, d\tilde{\mu}_i^s \tag{8.6}$$

where the sum is over all components in the solution except the solvent. Since the bulk of the solution and the interface are in equilibrium, the respective electrochemical potentials are equal. We can then eliminate the solvent terms from Eq. (8.5) with the aid of Eq. (8.6), and define the *surface excess* of species i through:

$$\Gamma_i = \Gamma_i^* - \frac{N_i^s}{N_0^s} \Gamma_0^* \tag{8.7}$$

These excess quantities are independent of the thickness chosen for the interface as long as it incorporates the region where the concentrations are different from those in the bulk; that is, it does not matter if one chooses too thick a region (see Problem 1). We cannot refer the surface concentrations of the metal particles M, M^{z+}, and e^- to the solution. Nevertheless we will drop the asterisk in their surface concentrations to simplify the writing; we will eliminate these quantities later. We can now rewrite the Gibbs adsorption equation in the form:

$$d\gamma = -\sum_i^{sol}{}' \Gamma_i \, d\tilde{\mu}_i^s - \Gamma_{M^{z+}} \, d\tilde{\mu}_{M^{z+}}^\sigma - \Gamma_e \, d\tilde{\mu}_e^\sigma - \Gamma_M \, d\mu_M^\sigma \tag{8.8}$$

[1] From here on we follow the classical derivation.

where the sum is over all species occurring in the solution except the solvent.

The metal ions M^{z+}, the atoms M, and the electrons at the interface are in equilibrium with the metal; so we may use the electrochemical potentials of these species in the metal instead of the interfacial quantities, and split them into the chemical part and the electrostatic part:

$$\begin{aligned} & -\Gamma_{M^{z+}} d\tilde{\mu}^{\sigma}_{M^{z+}} - \Gamma_e d\tilde{\mu}^{\sigma}_e - \Gamma_M d\mu^{\sigma}_M \\ &= -\Gamma_{M^{z+}} d\mu^m_{M^{z+}} - \Gamma_e d\mu^m_e - d\phi^m \left(z e_0 \Gamma_{M^{z+}} - e_0 \Gamma_e \right) - \Gamma_M d\mu^m_M \\ &= -\Gamma_{M^{z+}} d\mu^m_{M^{z+}} - \Gamma_e d\mu^m_e - \sigma d\phi^m - \Gamma_M d\mu^m_M \end{aligned} \tag{8.9}$$

where $\sigma = e_0(z\Gamma_{M^{z+}} - \Gamma_e)$ is the surface charge density due to the particles in common with the metal. Since the interface is electrically neutral, this must be balanced by the surface charge density due the ions in common with the solution:

$$\sigma = z e_0 \Gamma_{M^{z+}} - e_0 \Gamma_e = -\sum_j z_j e_0 \Gamma_j \tag{8.10}$$

where the sum is over all ionic species in the solution. Again we split the electrochemical potential into its chemical and electrostatic part: $\tilde{\mu}^s_j = \mu^s_j + z_j \phi^s$. On the metal side, the metal ions, atoms, and electrons are in equilibrium; hence: $\mu_M = \mu_{M^{z+}} + z\mu_e$. Substituting these relations into Eq. (8.8) gives:

$$d\gamma = -\sigma \; d(\phi^m - \phi^s) - \sum_j \Gamma_j \; d\mu^s_j - \sum_k \Gamma_k \; d\mu^s_k \tag{8.11}$$

The first sum is over all ionic species in the solution, the second sum over all neutral species except the metal atoms. For a pure metal the concentration of the metal atoms is constant; so the differential of the chemical potential of the metal atoms vanishes: $d\mu_M = 0$; we note in passing that complications can arise for amalgams, if the surface concentration of the metal atoms changes. All chemical potentials in Eq. (8.11) refer to the solution.

The difference $\phi^m - \phi^s$ in the inner potentials is not directly measurable; however, if the solution is in contact with a suitable reference electrode, its inner potential with respect to this electrode is fixed, and $d(\phi^m - \phi^s) = d\phi$, where ϕ is the electrode potential. The resulting equation is known as the *electrocapillary equation*:

$$d\gamma = -\sigma \; d\phi - \sum_j \Gamma_j \; d\mu^s_j - \sum_k \Gamma_k \; d\mu^s_k = -\sigma \; d\phi - \sum_i \Gamma_i \; d\mu^s_i \tag{8.12}$$

where we have combined the two sums into one, so that the sum is over all solution species except the solvent. The structure of this equation is worth noting: The intensive variables ϕ and μ refer to the two adjoining bulk phases, ϕ to the metal, and the chemical potentials to the solution; they can easily be measured and controlled. The variables σ and Γ refer to the interface.

The charge density σ deserves a special comment. Its definition is formal in the sense that from a thermodynamic point of view we know nothing about the actual distribution of the charge. It acquires its meaning only within a model in which the metal charge and the ionic charge form a double layer, with the metal charge forming an excess on the metal side of the interface, and the ionic charge an excess on the solution side.

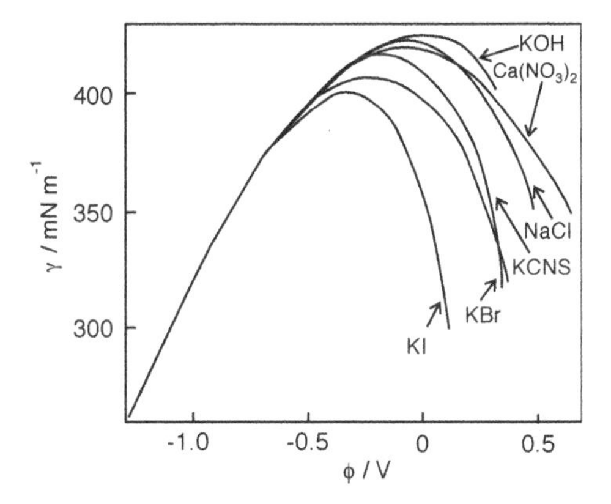

Fig. 8.2. Interfacial tension γ of a mercury electrode as a function of the electrode potential for 0.1 M aqueous solutions of several electrolytes at 18°C. The potential is given with respect to the pzc of a solution of KF. Data taken from [1].

If the interfacial tension γ can be measured, the surface charge density can be obtained by differentiation, which yields the *Lippmann equation*:

$$\sigma = -\left(\frac{\partial \gamma}{\partial \phi}\right)_{\mu_i} \tag{8.13}$$

This equation further indicates that the interfacial tension has an extremum at the pzc; differentiating again gives the differential interfacial capacity:

$$\left(\frac{\partial^2 \gamma}{\partial \phi^2}\right)_{\mu_i} = -C \tag{8.14}$$

Since C must be positive, this extremum is a maximum. Figure 8.2 shows a few examples of *electrocapillary curves*, in which the surface tension of a mercury electrode is measured as a function of the electrode potential at constant composition of the solution. At low potentials the various curves coincide, indicating that the cations are not specifically adsorbed. In contrast, the anions are adsorbed at higher potentials, so that the curves diverge. Note that with increasing adsorption the pzc is shifted to higher potentials. This corresponds to the increase of the work function caused by anion adsorption.

The electrocapillary equation (8.12) makes it possible to measure the surface excess of a species through:

$$\Gamma_i = -\left(\frac{\partial \gamma}{\partial \mu_i}\right)_{\mu_j \neq \mu_i} \tag{8.15}$$

If the species is neutral, its chemical potential μ_i can be varied by changing its concentration and hence its activity a_i: $d\mu_i = RT\ d\ln a_i$. In this case the determination of the surface excesses offers no difficulty in principle. However, if a species is charged, its concentration cannot be varied independently from that of a counterion, since the solution must be electrically neutral. To be specific, we consider the case of a 1–1 electrolyte composed of monovalent ions A^- and B^+. The electrocapillary equation then takes the form:

$$d\gamma = -\sigma\ d\phi - \Gamma_{A^-}\ d\mu_{A^-} - \Gamma_{B^+}\ d\mu_{B^+} \tag{8.16}$$

The two surface excesses are related through:

$$-\sigma = e_0\left(\Gamma_{B^+} - \Gamma_{A^-}\right) \tag{8.17}$$

since the charge on the metal must be balanced by the ionic charge at the interface. So we can rewrite Eq. (8.16) in the following form:

$$d\gamma = -\sigma\left(d\phi - \frac{1}{e_0}d\mu_{A^-}\right) - \Gamma_{B^+}\left(d\mu_{B^+} + d\mu_{A^-}\right) \tag{8.18}$$

The first term in parentheses has the following meaning: If a reference electrode is used whose potential is determined by a simple exchange reaction involving the anion A^-, the electrode potential ϕ_A with respect to this reference will depend on the concentration of the anion, and $d\phi_A = d\phi - d\mu_{A^-}/e_0$. The term $d\mu_{B^+} + d\mu_{A^-}$ denotes the change in the chemical potential of the uncharged species AB, and is determined by the change in the mean activity $2RT\ d\ln a_\pm$. Hence:

$$d\gamma = -\sigma\ d\phi_A - 2RT\ \Gamma_{B^+}\ d\ln a_\pm \tag{8.19}$$

and the surface excess of the cation can be determined through:

$$\Gamma_{B^+} = -\frac{1}{2RT}\left(\frac{\partial \gamma}{\partial \ln a_\pm}\right)_{\phi_A} \tag{8.20}$$

The surface excess of the anion is then obtained from the charge balance condition. Usually it is not practicable to use a reference electrode involving the anion. Instead, one uses a convenient reference electrode and measures the surface tension over a range of potentials and concentrations of the adsorbed species. Then one calculates the corresponding potential ϕ_A and determines the derivative in Eq. (8.20) numerically.

Such measurements require great precision. Figure 8.3 shows the surface excesses of a few ions. Instead of the surface excesses Γ, the corresponding excess charges $\sigma_\pm = zF\Gamma_\pm$ are shown. Note that the potassium cation is

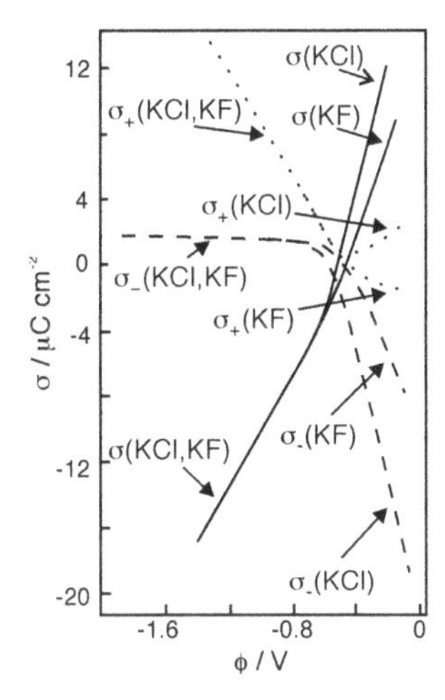

Fig. 8.3. Metal charge σ and ionic charges σ_+ and σ_- for a mercury electrode in contact with 0.1 M solutions of KF and KCl. Data taken from [1].

not specifically adsorbed. In contrast the Cl^- ion is strongly adsorbed at higher potentials. The high surface excess of the anion induces even a positive surface excess of the cation in the region positive of the pzc. The F^- ion is only weakly adsorbed, and for a solution of KF the surface excesses of both ions pass through zero near the pzc.

8.2 Solid electrodes

For solid electrodes the surface tension γ is the work done in forming a unit area of the metal by cleaving. One can also create new surface area by stretching, which gives rise the concepts of surface stress and strain. In fact, surface stress introduces an extra term into the Lippmann equation. However, this term is small and usually negligible; therefore we ignore it for the moment, and relegate the treatment of surface stress to a later section in this chapter.

It is practically impossible to measure γ for solid electrodes. However, in some applications one needs only the change in γ with certain parameters. For example, for the determination of the surface excess of a neutral organic species, one requires the change in the interfacial tension with the activity of the species. This can be measured if there is a reference potential ϕ_r at which the species is not adsorbed; the change in the interfacial tension is then referred to this potential. One proceeds in the following way [4]:

1. Determine the pzc of the electrode in the absence of the adsorbate; this can be done by finding the minimum of the interfacial capacity for a low concentration of the supporting electrolyte.

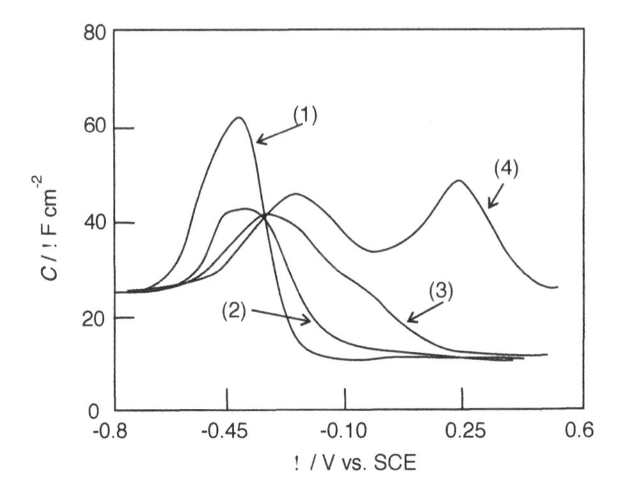

Fig. 8.4. Differential interfacial capacity for a Au(110) surface in contact with aqueous solutions containing 0.1 M $KClO_4$ and various amounts of pyridine. (1) no pyridine; (2) 3×10^{-5} M ; (3) 10^{-4} M; (4) 6×10^{-4} M pyridine. Data taken from [4].

2. Still in the absence of the adsorbate, measure the charge σ on the electrode over a range of electrode potentials ϕ (including ϕ_r) by stepping the potential from the pzc to ϕ and integrating over the current; alternatively one can obtain $\sigma(\phi)$ by measuring the interfacial capacity and integrating over ϕ.
3. The relative surface tension is then obtained by integration:

$$\Delta\gamma = \gamma(\phi) - \gamma(\phi_r) = \int_{\phi_r}^{\phi} \sigma q(\phi')\, d\phi' \tag{8.21}$$

4. This procedure is repeated for a range of concentrations c of the adsorbate, so that one obtains $\gamma(\phi, c) - \gamma(\phi_r)$. Since the adsorbate is not adsorbed at ϕ_r, the reference point $\gamma(\phi_r)$ is independent of the concentration of the adsorbate.
5. Denoting by $\gamma_0(\phi)$ the surface tension in the absence of the adsorbate, we obtain the surface excess from:

$$\Gamma = \frac{1}{RT}\left(\frac{\partial\left(\gamma(\phi, c) - \gamma_0(\phi)\right)}{\partial \ln(a)}\right)_{\phi} \tag{8.22}$$

Again, such measurements require great precision; therefore this method, though very exact in principle, is seldom used.

As an example we consider the adsorption of pyridine on Au(110) from a solution containing 0.1 M $KClO_4$ [4]. Figure 8.4 shows the differential capacity both in the absence and in the presence of various amounts of pyridine. Since the capacity curves coincide for potentials below about 0.7 V vs. SCE, the

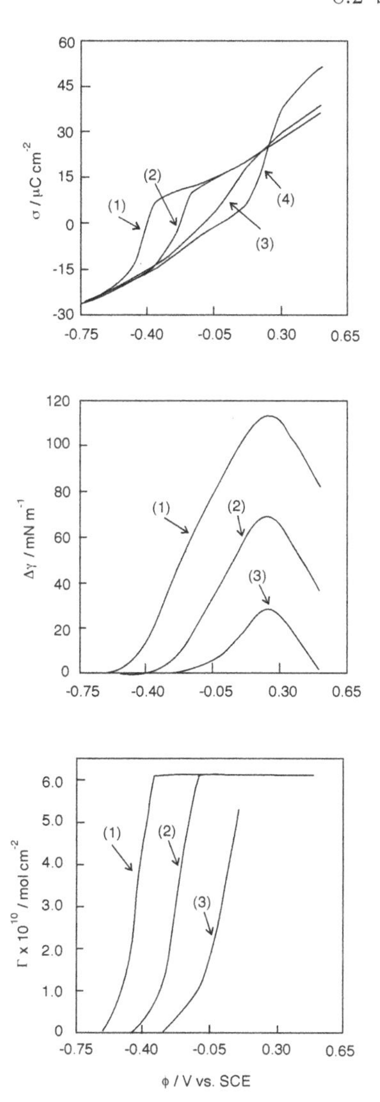

Fig. 8.5. Data required for the determination of the surface excess of pyridine on Au(110). *Top*: surface charge density obtained by integrating the capacity curves in Fig. 8.4. *Middle*: relative surface tension $\Delta\gamma$. *Bottom*: Surface excess of pyridine. Supporting electrolyte: 0.1 M $KClO_4$. (1) 6×10^{-4} M; (2) 3×10^{-5} M ; (3) 2×10^{-6} M pyridine; (4) no pyridine. Data taken from [4].

reference potential was chosen at -0.75 V. Integration with respect to the potential gives the surface charge density (see Fig. 8.5), another integration the relative surface tension. Finally, differentiation gives the surface excess.

8.3 Surface stress

For a solid, there are two different ways of creating new surface area: cleaving, which is a plastic deformation, and stretching, which is elastic. The energy change associated with cleaving is γdA_{p}, where p stands for plastic. For liquids, this is the only way of creating a new surface, Stretching can occur in various directions, so the surface strain ϵ associated with stretching as really a tensor. To simplify matters, we regard here only isotropic surfaces, like the (111) surfaces of fcc crystals, for which the tensor is diagonal, and the change in surface strain can simply be written as $d\epsilon = dA_{\mathrm{e}}/A$, where the index e stands for elastic; the general treatment can be found in the literature [5] – in fact, as far as we are aware, all electrochemical experiments have been performed on isotropic surfaces. The change in energy associated with a elastic deformation is $g\ dA_{\mathrm{e}}$, where g is the surface stress. Therefore, we have for the thermodynamic potential $X = \gamma A$:

$$\frac{\partial X}{\partial A_{\mathrm{e}}} = A\frac{\partial \gamma}{\partial A_{\mathrm{e}}} + \gamma = g \tag{8.23}$$

or on rearrangement:

$$A\frac{\partial \gamma}{\partial A_{\mathrm{e}}} = \frac{\partial \gamma}{\partial \epsilon} = g - \gamma \tag{8.24}$$

Adding this term to the electrocapillary equation gives:

$$d\gamma = -\sigma d\phi - \sum_i \Gamma_i d\mu_i + (g - \gamma)d\epsilon \tag{8.25}$$

In the case of a non-isotropic surface, where g and ϵ are tensors, the last term must be replaced by: $\sum_{n,m}(g_{nm} - \gamma\delta_{nm})\ d\epsilon_{nm}$. Equation (8.24) is known as Shuttleworth's equation and usually written in the form:

$$g = \gamma + \frac{\partial \gamma}{\partial \epsilon}, \quad \text{or more generally:} \quad g_{nm} = \gamma\delta_{nm} + \frac{\partial \gamma}{\partial \epsilon_{nm}} \tag{8.26}$$

For a liquid electrode, $\partial\gamma/\partial\epsilon = 0$, so that surface stress and tension are equal, and the last term in Eq. (8.25) vanishes.

In electrochemistry, the variation of the surface tension with the electrode potential is important. Experiments by Ibach et al. [6] and others have demonstrated that $\partial\epsilon/d\phi$ is small, and the last term in Eq. (8.25) is about a factor of 10^{-6} smaller than the first, and can therefore be neglected when calculating changes in the surface tension. Nevertheless, the study of the surface stress is an interesting topic in its own right. As an example, we show the variation of the surface stress on Au(111) with electrode potential (see Fig. 8.6). The dependence is significant, and sensitive to the surface structure. Thus, it is lower at an initially reconstructed surface than on the unreconstructed one. Also, g differs substantially from the surface tension, and does not have a maximum at the pzc, indicating that the second term in Shuttleworth's equation (8.26) is not negligible, and that the surface tension varies with the surface strain.

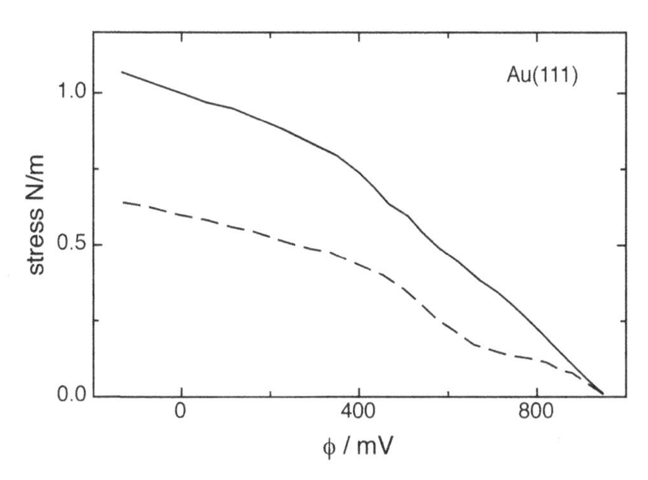

Fig. 8.6. Changes in the surface stress of a Au(111) electrode with potential (vs. SCE). The *lower curve* is for a surface that initially, at low potentials, is reconstructed. As the potential is scanned in the positive direction, the reconstruction is lifted. The *upper curve* is for the unreconstructed surface. Data taken from [6].

8.4 A note on the electrosorption valence

Here we deliver the proof, that the two definitions of the electrosorption valence are equivalent:

$$le_0 = \left(\frac{\partial \sigma}{\partial \Gamma_i}\right)_{\phi,\Gamma_j \neq \Gamma_i} = \left(\frac{\partial \mu_i^s}{\partial \phi}\right)_{\Gamma_i} \tag{8.27}$$

In order to prove this relation, we introduce an auxiliary quantity related to the surface tension γ:

$$Y = \gamma + \sum_j \Gamma_j \mu_j^s \tag{8.28}$$

where we have used the same notation as above. Using the Lippmann equation, the differential of Y is:

$$dY = \sum_j \mu_j^s \, d\Gamma_j - \sigma \, d\phi \tag{8.29}$$

which can be written as:

$$\left(\frac{\partial Y}{\partial \phi}\right)_{\Gamma_i} = -\sigma, \qquad \left(\frac{\partial Y}{\partial \Gamma_i}\right)_{\Gamma_j \neq \Gamma_i,\sigma} = \mu_i^s \tag{8.30}$$

Differentiating again gives:

$$le_0 = -\left(\frac{\partial \sigma}{\partial \Gamma_i}\right)_{\phi, \Gamma_j \neq \Gamma_i} = \frac{\partial Y}{\partial \phi \partial \Gamma_i} = \left(\frac{\partial \mu_i^s}{\partial \phi}\right)_{\Gamma_i} \tag{8.31}$$

where we have used the fact that Y is a proper function, and hence the order of differentiation does not matter. This is the desired relationship.

8.5 Potential of total zero charge

Finally, we want to clarify rarely-used variations of the concept of the potential of zero charge.[2] For simplicity we consider the case where a single species is specifically adsorbed with surface excess Γ. The surface charge density σ on the electrode then depends both on the electrode potential ϕ and on Γ. Therefore we can write for its differential:

$$d\sigma = \left(\frac{\partial \sigma}{\partial \phi}\right)_{\Gamma} d\phi + \left(\frac{\partial \sigma}{\partial \Gamma}\right)_{\phi} d\Gamma = \left(\frac{\partial \sigma}{\partial \phi}\right)_{\Gamma} d\phi + le_0 \, d\Gamma \tag{8.32}$$

where l is the electrosorption valence. If more than one species is adsorbed, the second term must be replaced by a sum over all adsorbates. In an experiment, generally both the potential ϕ and the coverage Γ change, and the corresponding change in σ, which has been defined as the electronic charge on the metal, is called the change in the *total charge density.* Changes in σ can be measured as the charge flowing into the electrode. The total charge in the interface is always zero. Conceptually, the countercharge to the electronic charge can be divided into the charge stored in the diffuse double layer and represented by the first term, and the charge stored in the adsorbate. Note that this division implies a specific model for the interface. The potential at which the total charge density σ vanishes is called the *potential of zero total charge.* It corresponds to the maximum of the surface tension, but in general not to the minimum of the Gouy–Chapman capacity.

Under favorable circumstances it may be possible to vary the potential without changing the surface excess Γ. In this case the double-layer capacitance, for sufficiently dilute solutions, has a minimum at the potential where the charge in the diffuse double-layer vanishes. The latter charge is sometimes called the free charge, and the potential where it vanishes the *potential of zero free charge.* We can express the same idea in terms of concepts familiar from surface science: An adsorbate changes the work function, and induces a similar change in the potential of zero charge, which is then called the potential of zero free charge. Both the potential of zero total charge and that of free charge depend on the adsorbate, and are not fundamental properties of the electrode material.

The potential of total zero charge is often invoked for the platinum-group metals [7], which are practically always covered by an adsorbate.

---

[2] We thank our colleagues Juan Feliu and Jacek Lipkowski for enlightening discussions on these concepts.

Problems

1. Show that the excess quantities defined in Eq. (8.7) are independent of the thickness chosen for the interface as long as the interface incorporates all of the regions where the concentrations are different from the bulk.
2. The Parsons function ξ is defined through: $\xi = \gamma + \sigma\phi$; it is the thermodynamic potential that has the charge density σ as the basic variable instead of the potential ϕ. Show that the surface excess of a species can be obtained through:

$$\Gamma_i = \frac{1}{RT}\left(\frac{\partial \xi}{\partial \ln a}\right)_\sigma \tag{8.33}$$

3. Prove Shuttleworth's equation directly by (a) first cleaving and then stretching a phase, (b) first stretching and then cleaving it, and comparing the results.

References

1. D.C. Grahame, *Chem. Revs.* **41** (1947) 441; D.C. Grahame and B.A. Soderberg, *J. Chem. Phys.* **22** (1954) 449.
2. R. Parsons, *Comprehensive Treatise of Electrochemistry*, Vol. I, edited by J. O'M. Bockris, B.E. Conway, and E. Yeager, Plenum Press, New York, NY, 1980.
3. H. Ibach, *Physics of Surfaces and Interfaces*, Springer, Berlin, Heidelberg, 2006.
4. J. Lipkowski and L. Stolberg, *Adsorption of Molecules at Electrodes*, edited by J. Lipkowski and P.N. Ross. VCH, New York, 1992.
5. R.G. Linford, *Chem. Rev.* **78** (1978) 81.
6. H. Ibach, C.E. Bach, M. Giesen, and A. Grossmann, *Surf. Sci.* **375** (1997) 107.
7. V. Climent, R. Gomez, and J.M. Feliu, *Electrochim. Acta* **45** (1999) 629.

9

Phenomenological treatment of electron-transfer reactions

9.1 Outer-sphere electron-transfer

Electron-transfer reactions are the simplest class of electrochemical reactions. They play a special role in that every electrochemical reaction involves at least one electron-transfer step. This is even true if the current across the electrochemical interface is carried by ions since, depending on the direction of the current, the ions must either be generated or discharged by an exchange of electrons with the surroundings.

In general electron-transfer reactions can be quite complicated, involving breaking or forming of chemical bonds, adsorption of at least one of the redox partners, or the presence of certain catalysts. Here we treat the simplest possible case, so-called *outer-sphere electron-transfer reactions*, in which from a chemist's point of view nothing happens but the exchange of one electron – as we shall see later, the simultaneous transfer of two or more electrons is highly unlikely. In the course of such a reaction, no bonds are broken or formed, the reactants are not specifically adsorbed, and catalysts play no role. If one of these conditions is not fulfilled, the reaction is said to proceed via an *inner-sphere* pathway. Unfortunately, there are not many examples for outer sphere reactions; here are two:

$$\begin{aligned} [\mathrm{Ru(NH_3)_6}]^{2+} &\rightleftharpoons [\mathrm{Ru(NH_3)_6}]^{3+} + \mathrm{e}^- \\ [\mathrm{Fe(H_2O)_6}]^{2+} &\rightleftharpoons [\mathrm{Fe(H_2O)_6}]^{3+} + \mathrm{e}^- \end{aligned} \qquad (9.1)$$

In aqueous solutions these reactions seem to proceed via an outer-sphere mechanism on most metals. Typically such reactions involve metal ions surrounded by inert ligands, which prevent adsorption. Note that the last example reacts via an outer-sphere pathway only if trace impurities of halide ions are carefully removed from the solution; otherwise it is catalyzed by these ions.

W. Schmickler, E. Santos, *Interfacial Electrochemistry*, 2nd ed.,
DOI 10.1007/978-3-642-04937-8_9, © Springer-Verlag Berlin Heidelberg 2010

9.2 The Butler–Volmer equation

In this chapter we treat electron-transfer reactions from a macroscopic point of view using concepts familiar from chemical kinetics. The overall rate v of an electrochemical reaction is the difference between the rates of oxidation (the *anodic reaction*) and reduction (the *cathodic reaction*); it is customary to denote the anodic reaction, and the current associated with it, as positive:

$$v = k_{\rm ox} c^s_{\rm red} - k_{\rm red} c^s_{\rm ox} \tag{9.2}$$

where $c^s_{\rm red}, c^s_{\rm ox}$ denote the surface concentrations of the reduced and oxidized species, and $k_{\rm ox}$ and $k_{\rm red}$ are the rate constants. Using absolute rate theory, the latter can be written in the form:

$$\begin{aligned} k_{\rm ox} &= A \exp\left(-\frac{\Delta G^\dagger_{\rm ox}(\phi)}{RT}\right) \\ k_{\rm red} &= A \exp\left(-\frac{\Delta G^\dagger_{\rm red}(\phi)}{RT}\right) \end{aligned} \tag{9.3}$$

The phenomenological treatment assumes that the Gibbs energies of activation $G_{\rm ox}$ and $G_{\rm red}$ depend on the electrode potential ϕ, but that the pre-exponential factor A does not. We expand the energy of activation about the standard equilibrium potential ϕ_{00} of the redox reaction; keeping terms up to first order, we obtain for the anodic reaction:

$$\Delta G^\dagger_{\rm ox}(\phi) = \Delta G^\dagger_{\rm ox}(\phi_{00}) - \alpha F(\phi - \phi_{00}), \quad \text{with} \quad \alpha = -\frac{1}{F} \left.\frac{\partial \Delta G^\dagger_{\rm ox}}{\partial \phi}\right|_{\phi_{00}} \tag{9.4}$$

The quantity α is the *anodic transfer coefficient*; the factor $1/F$ was introduced, because $F\phi$ is the electrostatic contribution to the molar Gibbs energy, and the sign was chosen such that α is positive – obviously an increase in the electrode potential makes the anodic reaction go faster, and decreases the corresponding energy of activation. Note that α is dimensionless. For the cathodic reaction:

$$\Delta G^\dagger_{\rm red}(\phi) = \Delta G^\dagger_{\rm red}(\phi_{00}) + \beta F(\phi - \phi_{00}), \quad \text{with} \quad \beta = \frac{1}{F} \left.\frac{\partial \Delta G^\dagger_{\rm red}}{\partial \phi}\right|_{\phi_{00}} \tag{9.5}$$

where the *cathodic transfer coefficient* β is also positive. One would expect that higher terms in the expansion of the Gibbs energy of activation will become important at potentials far from the standard equilibrium potential ϕ_{00}; we will return to this point in the next chapter. The Gibbs energies of activation are related by:

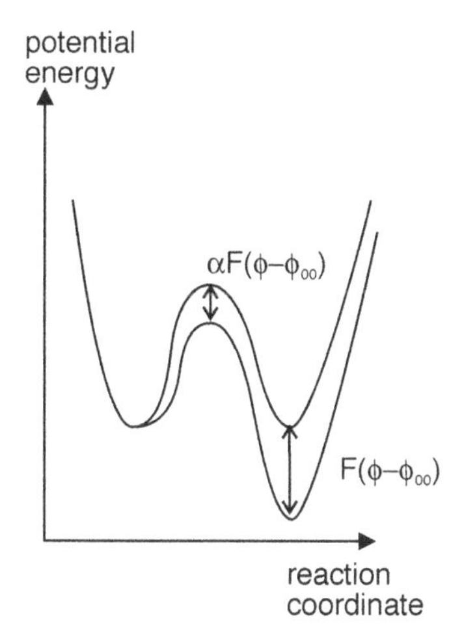

Fig. 9.1. Potential energy curves for an outer-sphere reaction; the *upper curve* is for the standard equilibrium potential ϕ_{00}; the *lower curve* for $\phi > \phi_{00}$.

$$\Delta G^{\dagger}_{\text{ox}}(\phi) - \Delta G^{\dagger}_{\text{red}}(\phi) = G_{\text{ox}} - G_{\text{red}} \tag{9.6}$$

to the molar Gibbs energies G_{ox} and G_{red} of the oxidized and reduced state; in particular:

$$\Delta G^{\dagger}_{\text{ox}}(\phi_{00}) = \Delta G^{\dagger}_{\text{red}}(\phi_{00}) = \Delta G^{\dagger}_{00} \tag{9.7}$$

When the electrode potential is changed from ϕ_{00} to a value ϕ, the Gibbs energy of the electrons on the electrode is lowered by an amount $-F(\phi - \phi_{00})$, and so is the energy of the oxidized state. If the reactants are so far from the metal surface that their electrostatic potentials are unchanged when the electrode potential is varied, then the Gibbs energy of the reaction is also changed by $-F(\phi - \phi_{00})$. This condition is generally fulfilled for outer-sphere reactions in the presence of a high concentration of an inert electrolyte which screens the electrode potential; it is not fulfilled when the reactants are adsorbed as in inner-sphere reactions. When it is fulfilled we have:

$$\Delta G^{\dagger}_{\text{ox}}(\phi) - \Delta G^{\dagger}_{\text{red}}(\phi) = -F(\phi - \phi_{00}) \tag{9.8}$$

By differentiation we obtain for the sume of the two transfer coefficients the relation:

$$\alpha + \beta = 1 \tag{9.9}$$

Since both coefficients are positive, they lie between zero and one; we can generally expect a value near 1/2 unless the reaction is strongly unsymmetrical.

The transfer coefficients have a simple geometrical interpretation. In a one-dimensional picture we can plot the potential energy of the system as a function of a generalized reaction coordinate (see Fig. 9.1). The reduced and the oxidized states are separated by an energy barrier. Changing the electrode potential by an amount $(\phi - \phi_{00})$ changes the molar Gibbs energy of the oxidized state by $-F(\phi - \phi_{00})$; the Gibbs energy of the transition state located at the maximum will generally change by a fraction $-\alpha F(\phi - \phi_{00})$, where $0 < \alpha < 1$. The relation $\alpha + \beta = 1$ is easily derived from this picture.

The current density j associated with the reaction is simply $j = Fv$. Combining Eqs. (9.2) to (9.5) and (9.9) gives the *Butler–Volmer* equation [1, 2] in the form:

$$j = Fk_0 c^s_{\text{red}} \exp \frac{\alpha F(\phi - \phi_{00})}{RT} - Fk_0 c^s_{\text{ox}} \exp\left(-\frac{(1-\alpha)F(\phi - \phi_{00})}{RT}\right) \tag{9.10}$$

where

$$k_0 = A \exp\left(-\frac{\Delta G^\dagger(\phi_{00})}{RT}\right) \tag{9.11}$$

Using the Nernst equation:

$$\phi_0 = \phi_{00} + \frac{RT}{F} \ln \frac{c^s_{\text{ox}}}{c^s_{\text{red}}} \tag{9.12}$$

for the equilibrium potential ϕ_0, and introducing the *overpotential* $\eta = \phi - \phi_0$, which is the deviation from the equilibrium potential, we rewrite the Butler–Volmer equation in the form:

$$j = j_0 \left[\exp \frac{\alpha F \eta}{RT} - \exp\left(-\frac{(1-\alpha)F\eta}{RT}\right)\right] \tag{9.13}$$

where

$$j_0 = Fk_0 (c^s_{\text{red}})^{(1-\alpha)} (c^s_{\text{ox}})^\alpha \tag{9.14}$$

is the *exchange current density.* At the equilibrium potential the anodic and cathodic current both have the magnitude j_0 but opposite sign, thus cancelling each other. The exchange current density for unit surface concentration of the reactants is the *standard exchange current density* $j_{00} = Fk_0$, which is a measure of the reaction rate at the standard equilibrium potential.

According to the Butler–Volmer law, the rates of simple electron-transfer reactions follow a particularly simple law. Both the anodic and the cathodic current densities depend exponentially on the overpotential η (see Fig. 9.2.) For large absolute values of η, one of the two partial currents dominates, and a plot of $\ln |j|$ – or of $\log_{10} |j|$ – versus η, a so-called *Tafel plot* [3] (see Fig. 9.3), yields a straight line in this region. From its slope and intercept the transfer coefficient and the exchange current density can be obtained. These two quantities completely determine the current-potential curve.

For small overpotentials, in the range $|F\eta| \ll RT$, the Butler–Volmer equation can be linearized by expanding the exponentials:

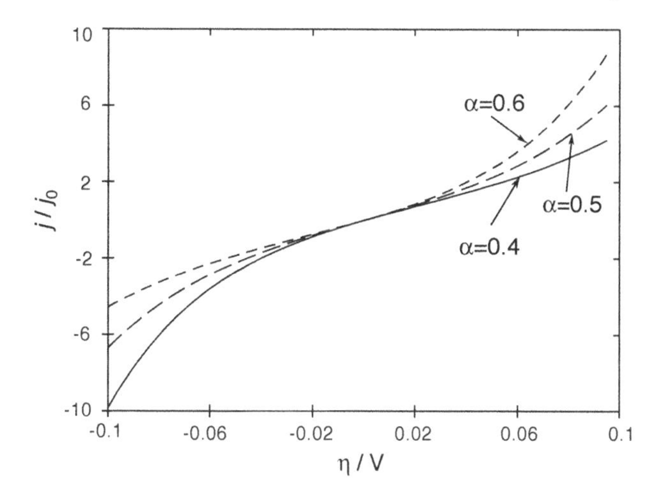

Fig. 9.2. Current-potential curves according to the Butler–Volmer equation.

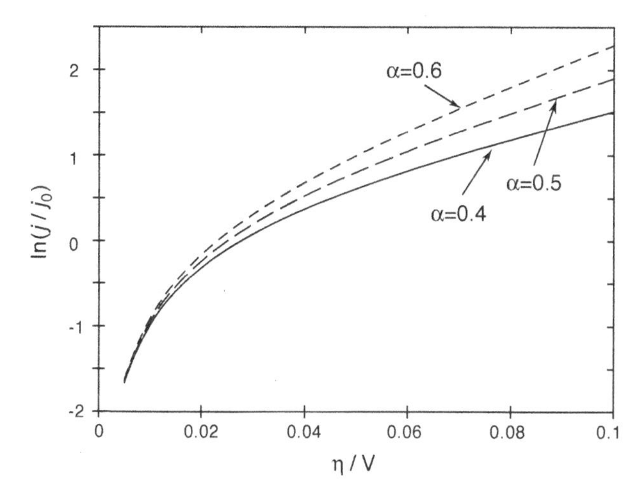

Fig. 9.3. Tafel plot for the anodic current density of an outer-sphere reaction.

$$j = j_0 \frac{F\eta}{RT} \tag{9.15}$$

The quantity $\eta/j = RT/j_0F$ is called the *charge-transfer resistance*. Note that the transfer coefficient does not appear in the current-voltage relation for small overpotentials, and hence cannot be determined from measurements at small deviations from equilibrium, they give the exchange current density only. However, α can be obtained by varying the surface concentrations, measuring the exchange current density, and using Eq. (9.14). We will discuss a few examples of outer-sphere electron-transfer reactions in Chap. 12.

We conclude these phenomenological considerations with a few remarks:

1. The transfer coefficient is equivalent to the Broenstedt coefficient well known from ordinary chemical kinetics. Both describe the change in the energy of activation with the Gibbs energy of the reaction.
2. The transfer coefficient α has a dual role: (1) It determines the dependence of the current on the electrode potential. (2) It gives the variation of the Gibbs energy of activation with potential, and hence affects the temperature dependence of the current. If an experimental value for α is obtained from current-potential curves, its value should be independent of temperature. A small temperature dependence may arise from quantum effects (not treated here), but a strong dependence is not compatible with an outer-sphere mechanism.
3. For small overpotentials the linear approximations of Eqs. (9.4) and (9.5) should be sufficient, but at high overpotentials higher-order terms are expected to contribute.
4. The transfer coefficient determines the symmetry – or lack thereof – of the current-potential curves; they are symmetric for $\alpha = 1/2$. For this reason the transfer coefficient is also known as the *symmetry factor*.
5. The surface concentrations are generally not known, and may vary with time as the reaction proceeds. One way to circumvent this problem is to work under conditions of controlled convection, so that the surface concentrations can be calculated from the bulk concentrations. Another technique consists in the use of potential or current pulses, which allows an extrapolation back to the time of the onset of the pulse when surface and bulk concentrations are equal. These techniques will be discussed in detail in Chaps. 19 and 20.
6. Inner-sphere electron-transfer reactions are not expected to obey the Butler–Volmer equation. In these reactions the breaking or formation of a bond, or an adsorption step, may be rate determining. When the reactant is adsorbed on the metal surface, the electrostatic potential that it experiences must change appreciably when the electrode potential is varied.

9.3 Double-layer corrections

When the concentration of the inert electrolyte is low, the electrostatic potential at the reaction site differs from that in the bulk and changes with the applied potential. This results in two effects [4]:

1. The surface concentrations $c^s_{\rm ox}$ and $c^s_{\rm red}$ differ from those in the bulk even if the surface region and the bulk are in equilibrium. Using the same arguments as in the Gouy–Chapman theory, the surface concentration c^s of a species with charge number z is:

$$c^s = c_0 \exp\left(-\frac{ze_0\phi_2}{kT}\right) \tag{9.16}$$

where c_0 is the bulk concentration, ϕ_2 the potential at the reaction site, and the potential in the bulk of the solution has been set to zero.

2. On application of an overpotential η, the Gibbs energy of the electron-transfer step changes by $e_0[\eta - \Delta\phi_2(\eta)]$, where $\Delta\phi_2(\eta)$ is the corresponding change in the potential ϕ_2 at the reaction site. Consequently, η must be replaced by $[\eta - \Delta\phi_2(\eta)]$ in the Butler–Volmer equation (9.13).

These modifications are known as the *Frumkin double-layer corrections.* They are useful when the electrolyte concentration is sufficiently low, so that ϕ_2 can be calculated from Gouy–Chapman theory, and the uncertainty in the position of the reaction site is unimportant. Whenever possible, kinetic investigations should be carried out with a high concentration of supporting electrolyte, so that double-layer corrections can be avoided.

9.4 A note on inner-sphere reactions

There is no general law for the current-potential characteristics of inner-sphere reactions. Depending on the system under consideration, various reaction steps can determine the overall rate: adsorption of the reacting species, an electron-transfer step, a preceding chemical reaction, coadsorption of a catalyst. If the rate-determining step is an outer-sphere reaction, the current will obey the Butler–Volmer equation. A similar equation may hold if an inner-sphere electron transfer, for example, from an adsorbed species to the metal, determines the rate. In this case, application of an overpotential η changes the Gibbs energy of this step only by a fraction of $F\eta$; furthermore, the concentration of the adsorbed species will change with η. These effects may result in phenomenological equations of the form:

$$k_{ox} = k_0 \exp\frac{\alpha F\eta}{RT}, \quad k_{red} = k_0 \exp\left(-\frac{\beta F\eta}{RT}\right) \tag{9.17}$$

with *apparent transfer coefficients* α and β, but α and β may depend on temperature.

If the rate-determining step is the adsorption of an ion, the reaction obeys the laws for ion-transfer reactions (see Chap. 13), and again a Butler–Volmer-type law will hold.

Problems

1. Derive Eq. (9.13) from Eqs. (9.10) and (9.12).

2. The reduced species of an outer-sphere electron-transfer reaction is generated by a chemical reaction of the form:

$$A \rightleftharpoons \text{red}$$

Denote the forward and backward rate constants of this reaction by k_a and k_b. When the reaction proceeds under stationary conditions, the rates of the chemical and of the electron-transfer reaction are equal. Derive the current-potential relationship for this case. Assume that the concentrations of A and of the oxidized species are constant.

3. The Gibbs energy of activation in Eq. (9.4) can be split into an enthalpy and an entropy term: $\Delta G^{\dagger}_{\text{ox}} = \Delta H^{\dagger}_{\text{ox}} - T\,\Delta S^{\dagger}_{\text{ox}}$. Define two transfer coefficients

$$\alpha_H = -\frac{1}{F}\frac{\partial \Delta H^{\dagger}_{\text{ox}}}{\partial(\phi - \phi_{00})}, \quad \alpha_S = \frac{1}{F}\frac{\partial \Delta S^{\dagger}_{\text{ox}}}{\partial(\phi - \phi_{00})}$$

and derive the corresponding current-potential relations. Note: For outer-sphere electron-transfer reactions α_S seems to be negligible; it has, however, been used to explain a temperature dependence of the apparent transfer coefficients in some inner-sphere reactions.

References

1. J.A. Butler, *Trans. Faraday Soc.* **19** (1924) 729.
2. T. Erdey-Gruz and M. Volmer, *Z. Physik. Chem.* **150** (1930) 203.
3. J. Tafel, *Z. Physik. Chem.* **50** (1905) 641
4. A.N. Frumkin, *Z. Physik. Chem.* **164A** (1933) 121.

10

Theoretical considerations of electron-transfer reactions

10.1 Qualitative aspects

Chemical and electrochemical reactions in condensed phases are generally quite complex processes, but outer-sphere electron-transfer reactions are sufficiently simple that we have reached a fair understanding of them in terms of microscopic concepts. In this chapter we give a simple derivation of a semi-classical theory of outer-sphere electron-transfer reactions, which was first systematically developed by Marcus [1] and Hush [2] in a series of papers. Several of the concepts that we develop here play also a role in electrocatalysis.

We begin with qualitative considerations. During the course of an outer-sphere electron-transfer reaction, the reactants get very close, up to a few Ångstroms, to the electrode surface. Electrons can tunnel over such a short distance, and the reaction would be very fast if nothing happened but the transfer of an electron. In fact, outer-sphere reactions are fast, but they have a measurable rate, and an energy of activation of typically 0.2–0.4 eV, since electron transfer is accompanied by reorganization processes of atoms and molecules that require thermal activation. While the reacting complex often has the same or similar structure in the oxidized and reduced form, metal-ligand bonds are typically shorter in the complex with the higher charge, which is also more strongly solvated. So the reaction is accompanied by a reorganization of both the complex, or *inner sphere*, and the solvation sheath, or *outer sphere* (see Fig. 10.1). These processes require an energy of activation and slow the reaction down.

A natural question is: In which temporal order do the reorganization processes and the proper electron transfer take place? The answer is given by the Frank–Condon principle, which in this context states: First the heavy particles of the inner and outer sphere must assume a suitable intermediate configuration, then the electron is exchanged isoenergetically, and finally the system relaxes to its new equilibrium configuration. A simple illustration is given in Fig. 10.2, where we have drawn potential energy surfaces for the reduced

W. Schmickler, E. Santos, *Interfacial Electrochemistry*, 2nd ed.,
DOI 10.1007/978-3-642-04937-8_10, © Springer-Verlag Berlin Heidelberg 2010

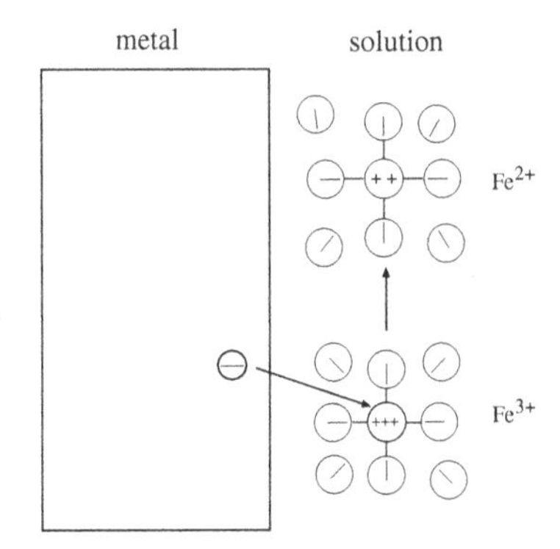

Fig. 10.1. Reorganization of *inner* and *outer* sphere during an electron-transfer reaction.

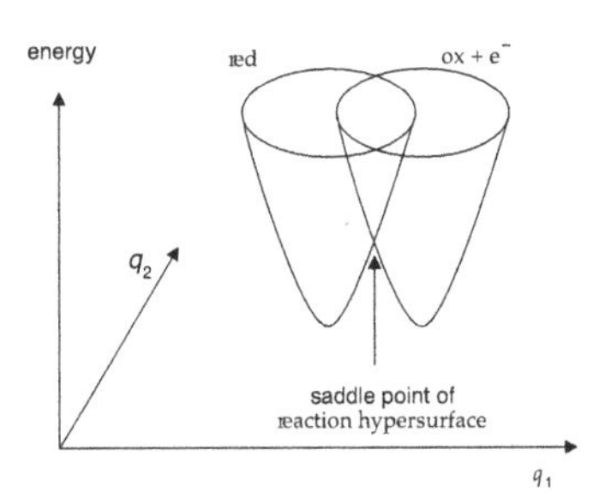

Fig. 10.2. Schematic diagram of the potential energy surfaces for the reduced and the oxidized state.

and the oxidized state as a function of two generalized reaction coordinates representing the positions of particles in the inner and outer sphere. During the course of an oxidation reaction, the system first moves along the surface for the reduced state till it reaches a crossing point with the surface for the oxidized state; at this configuration the electron can be transferred, and then the system moves to its new equilibrium position. Generally, the reaction will proceed via the saddle point of the intersection, since such transitions require the smallest energy of activation. The same diagram can also be used to illustrate the concept of *adiabaticity*: If the electron transfer takes place every time that the system is on the intersection surface, we speak of an *adiabatic*, otherwise of a *nonadiabatic* reaction. Of course, Fig. 10.2 is highly simplified: In reality, we must plot the potential energy as a function of the positions of all the heavy particles involved, so that we obtain multidimensional potential energy surfaces. Fortunately, for most purposes the dimensionality does not matter, and a one-dimensional model, which we will present below, suffices.

10.2 Harmonic oscillator with linear coupling

In an electron transfer reaction, we distinguish two different electronic states, an initial i and a final state f, which interact differently with the surroundings. The simplest model for a mode which interacts with the electron transfer is a harmonic oscillator. This amounts to expanding the energy of that mode about its equilibrium position and keeping terms up to second order. Thus, we may write for the energy of one such mode in the initial state:

$$E_i = \frac{1}{2} m\omega^2 x^2 \quad \text{or} \quad E_i = \frac{1}{2}\alpha x^2 \tag{10.1}$$

The former notation is natural for a real vibration, the latter for a more general case like a solvent mode. To first order, the interaction with the transferring electron is linear, i.e. proportional to x, and its strength is determined by a coupling constant g. It is convenient to define g in such a way that we may write for the final state:

$$E_f = \frac{1}{2}\alpha x^2 + \alpha g x \tag{10.2}$$

This can be rewritten as:

$$E_f = \frac{1}{2}\alpha (x+g)^2 - \lambda \quad \text{with} \quad \lambda = \frac{1}{2}\alpha g^2 \tag{10.3}$$

Thus, the origin of the harmonic oscillator has been shifted to $x = -g$, and its energy been lowered by an amount λ. Later we shall identify λ with the contribution of this mode to the energy of reorganization of the electron transfer reaction. It is convenient to simplify the notation by introducing a normalized coordinate $q = x/g$. This results in:

$$E_i = \lambda q^2 \qquad E_f = \lambda q^2 + 2\lambda q = \lambda\left(q^2 + 2q\right) \tag{10.4}$$

In general, there is also a change in electronic energy ΔE_e between final and initial state. If only a single mode is reorganized, the total change in energy between the equilibria of the final and initial states is:

$$\Delta E = \Delta G = \Delta E_e - \lambda \tag{10.5}$$

This change in energy may be identified with the change in the Gibbs free energy, because in this model there is no change in the entropy. If several modes couple, λ is the sum over all the contributions to the energy of reorganization. In the electrochemical case, we may write: $\Delta G = -e_0\eta$, where η is the overpotential. With these preliminaries, we are ready to consider electron transfer in a systematic manner.

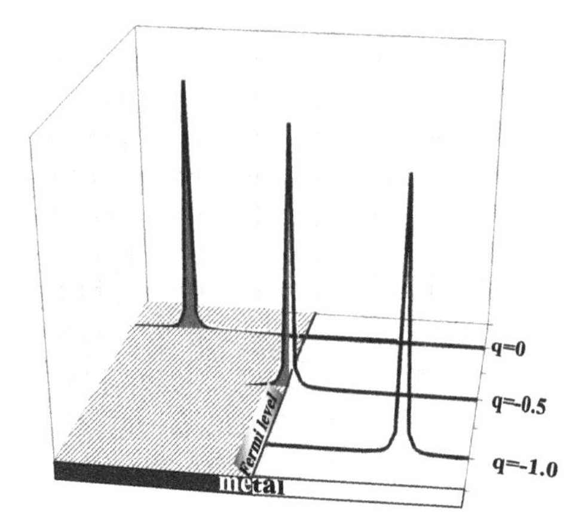

Fig. 10.3. Density of states of a reactant undergoing reaction 10.6 in the initial, the activated, and the final state, which, at equilibrium, in turn correspond to values of $q = 0$, $q = -1/2$, and $q = -1$ of the solvent coordinate.

10.3 Adiabatic electron transfer

Outer sphere electron reactions at bare metal electrodes are usually adiabatic. In this case the electron exchange between the reactant and the metal is faster than the motion of the inner or outer sphere modes, and the system is always in electronic equilibrium. We present a simplified version of the theory proposed by one of us [3].

To be specific, we consider an electron transfer of the type:

$$A \rightarrow B^+ + e^- \tag{10.6}$$

Before the electron transfer, the valence level of A is filled and lies below the Fermi level. As discussed in Chap. 6 for an adsorbed species, the interaction of the reactant with the metal broadens the valence level. It is therefore described by a density of states (DOS) of a certain width Δ; however, since the interaction of a non-adsorbed species is quite weak, the width Δ is small, typically of the order of 10^{-3} eV. In the final state, this valence level is empty and thus lies above the Fermi level (see Fig. 10.3). The ion B^+ interacts with the solvent, and the concomitant solvation energy makes it stable; without solvation, taking an energy from the reactant would only cost energy and would not result in a stable reactant. Electron transfer proceeds in the following way: A thermal fluctuation of the solvent lifts the valence level of A above the Fermi level, so that it is emptied. Then solvation sets in and lifts the valence level to its final state. The energy of activation is the energy required to lift the valence level from its initial state to the Fermi level.

For quantitative purposes we need expressions for the electronic energy and the solvation. For the former, we introduce the occupation n of the valence orbital, which is unity when the orbital lies below the Fermi level, and zero when it lies above. Neglecting the small finite width Δ for the moment, we write the electronic energy simply as $\epsilon_a n$, where ϵ_a is the energy of the valence level in the absence of the solvent.

For a full description of the reorganization of the outer and inner sphere modes, we would really have to consider the multidimensional potential energy surfaces sketched in Fig. 10.2. Thus, we would have two different surfaces for the initial and the final states, each with a different minimum corresponding to the equilibrium respective configuration. Fortunately, as long as only classical modes are reorganized, a one-dimensional model suffices, and gives the same results as a multi-dimensional one; the proof for this is relegated to Sect. 10.6. So, we consider a single reaction, or solvent, coordinate q, which in terms of Fig. 10.2 passes from the minimum of the initial surface via the saddle point to the minimum of the second surface. We can normalize this coordinate such that the resulting expression is as simple as possible. As a consequence, the calculated energy takes on the meaning of a free energy, since the other degrees of freedom that are not considered, have effectively been averaged out.

To simplify matters further, we follow the original papers of Marcus and Hush and use the harmonic approximation. Thus, in the initial state the energy of solvent[1] is represented by a simple parabola. As shown in the previous section, we can conveniently write this as λq^2, where λ is called the energy of reorganization of the reaction. When the reactant is ionized, it interacts with the solvent. To first order, this interaction is proportional to the charge on the ion, and will depend of the state of the solvent given by q; it can be written in the form $2\lambda q(1-n)$. That this expression has this simple form, is a consequence of the convenient normalization. In terms of the notation of the previous section, the term $(1-n)$ simply switches between initial and final state. Thus, the energy of the solvent is:

$$E_{\text{sol}} = \lambda q^2 + 2\lambda q(1-n) \tag{10.7}$$

In the initial state, when $n = 1$, the energy is a parabola with minimum at $q = 0$, and in the final state, when $n = 0$, a parabola with minimum at $q = -1$. Also, identifying $(1-n)$ with the charge number, Eq. (10.7) suggest an intuitive interpretation of the solvent coordinate: When the state of the solvent is characterized by q, it would be in equilibrium with a reactant of charge number $-q$.

Adding the electronic and the solvent terms gives the total energy as a function of the solvent coordinate q:

$$E(q) = \lambda q^2 + 2\lambda q + (\epsilon_a - 2\lambda q)n + \frac{\Delta}{2\pi} \ln \frac{(\epsilon_a - 2\lambda q)^2 + \Delta^2}{\epsilon_a^2 + \Delta^2} \tag{10.8}$$

[1] Really this includes the inner sphere as well, but for brevity we shall subsume it under the general term solvent

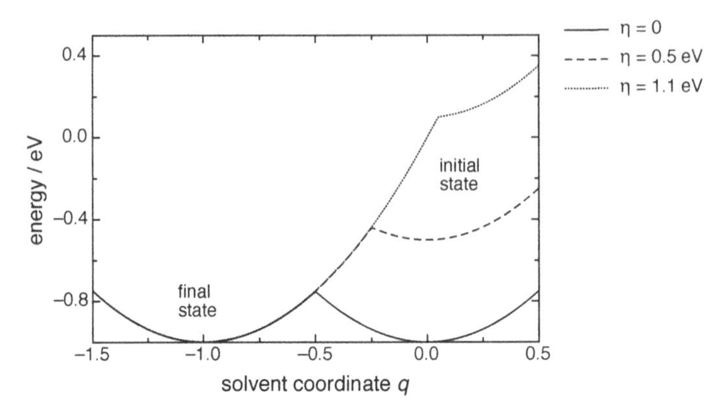

Fig. 10.4. Free energy curves for three different overpotentials. The energy of reorganization was taken as $\lambda = 1.0$ eV.

The last term is a small correction caused by the finite width Δ of the density of states. It can be neglected in most circumstances, but is qualitatively important at high overpotentials. The energy in the initial state, with $q = 0,\ n = 1$ is simply ϵ_a; in the final state, with $q = -1,\ n = 0$, it is $-\lambda$. Changing the electrode potential changes the level ϵ_a with respect to the Fermi level of the metal, which we take to be $E_F = 0$ as always. For $\epsilon_a = -\lambda$ the reaction is in equilibrium, so that we can write: $\epsilon_a = -\lambda + e_0\eta$, where η is the overpotential.

The energy as a function of the solvent coordinate q can easily be plotted from Eq. (10.8); a few examples for different overpotentials are shown in Fig. 10.4. Neglecting the last term, the formula gives two different parabolas for $n = 1$ and $n = 0$; for each value of q, the lower of the two energies is the correct adiabatic energy. For $|\eta| < \lambda$ the crossing of the two parabolas gives the activated state with coordinate q_s and a corresponding energy of activation, taken with respect to the initial energy $\epsilon_a = -\lambda + e_0\eta$:

$$q_s = \frac{-\lambda - e_0\eta}{2\lambda} \qquad E_{\text{act}} = \frac{(\lambda - e_0\eta)^2}{4\lambda} \qquad \text{for } |\eta| < \lambda \tag{10.9}$$

If we replace $-e_0\eta$ by the free reaction energy ΔG, the formula for the energy of activation is identical to that derived by Marcus and Hush for homogeneous electron transfer. For $\eta > \lambda$ the initial state is no longer stable, and electron transfer is immediate and without activation (see Fig. 10.4). The resulting energy of activation is shown in Fig. 10.5; it starts from a value of $E_{\text{act}} = \lambda/4$ at equilibrium and vanishes for $\eta \geq \lambda$. In contrast, the Marcus–Hush formula predicts an increase of the energy of activation with η for $\eta > \lambda$. For homogenous electron transfer, a decrease of the rate constant for very large values of $-\Delta G$ has indeed been observed. But this does not occur at metal electrodes, neither in the adiabatic nor in the non-adiabatic case, which will be treated in the next section.

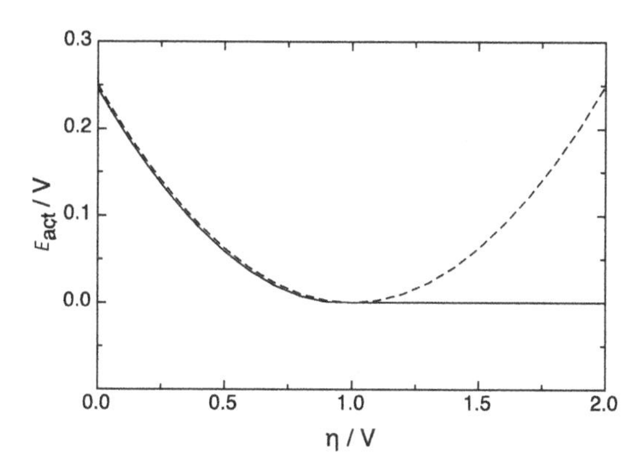

Fig. 10.5. Energy of activation as a function of the overpotential. *Full curve*: adiabatic theory; *dotted curve*: Marcus–Hush formula. $\lambda = 1.0$ eV.

At equilibrium, Eq. (10.9) predicts a transfer coefficient of $\alpha = 1/2$: indeed, experimental values for outer sphere reaction always lie close to this value. With increasing overpotential, α decreases, and for $\eta > \lambda$ it vanishes, as does the energy of activation, and the current becomes constant. The full curve will be displayed Fig. 10.6 in the next section for non-adiabatic transfer, where it has exactly the same shape.

In order to obtain the rate constant from the energy of activation, we require an estimate of the pre-exponential factor. As long as the reaction is adiabatic, it is independent of the strength of the interaction between the reactant and the electrode, but is solely governed by the dynamics of the reorganization. For water, typical times for the reorientation are of the order $10^{-12}-10^{-11}$ s, and the typical frequency is the inverse of this time. Adiabatic electron transfer requires the reactant to be no further than about $5-10$ Å from the electrode. Converting the bulk concentration to the number of particles in this range, we obtain a rough estimate of $A \approx 10^3$ cm s^{-1}, a number that is quite compatible with experimental data for fast outer sphere reactions (see Chap. 12). So the rate constant for the oxidation is now defined; the rate for the reverse reaction is obtained from the usual relation:

$$k_{\rm red} = \exp \frac{-e_0\eta}{kT}\, k_{\rm ox} \tag{10.10}$$

Finally we note, that larger values of the interaction Δ lead to a significant reduction of the energy of activation. We shall return to this effect when we consider electrocatalysis.

10.4 Non-adiabatic electron-transfer reactions

For non-adiabatic reactions the electronic interaction is much weaker, so that the system can pass the saddle point without an electron transfer, and subsequently return to its initial state. Therefore the interaction strength will enter into the pre-exponential factor. Also, we now have to consider into which electronic level the electron is actually transferred, a question that makes no sense for adiabatic reactions, since reactant and electrode simply share their electrons.

The basic ideas about solvent reorganization remain valid, but the equations for the energy require a few modifications. Firstly, in Eq. (10.8) we drop the last term, since Δ is even smaller than before. Further, we have to consider to which electronic level on the metal the electron is being transferred. We denote this energy by ϵ and, as always, measure it with respect to the Fermi level. The free energy of the reaction is reduced by ϵ, which has the same effect as replacing ϵ_a by $\epsilon_a - \epsilon$. Therefore we obtain for the free energy curves for transitions to a particular level ϵ:

$$E(q, \epsilon) = \lambda q^2 + 2\lambda q + (\epsilon_a - \epsilon - 2\lambda q)n \tag{10.11}$$

Again, we obtain two different parabolas for $n = 1$ and $n = 0$, but now we cannot argue, that the system is always on the surface with the lower energy, since the reaction is no longer adiabatic. Therefore, each parabola describes a redox state, and the energy of activation is obtained by calculating the intersection point. Equation (10.9) is now valid for all η, but again we have to replace ϵ_a by $\epsilon_a - \epsilon$:

$$E_{\text{act}}(\epsilon) = \frac{(\lambda + \epsilon - e_0\eta)^2}{4\lambda} \tag{10.12}$$

From this we can define a rate constant for the contribution that passes to an energy ϵ; bearing in mind that the transition can only take place if there is an empty level of energy ϵ on the electrode, we write:

$$k_{\text{ox}}(\epsilon) \propto [1 - f(\epsilon)] \exp - \frac{(\lambda + \epsilon - e_0\eta)^2}{4\lambda kT} \tag{10.13}$$

where $f(\epsilon)$ is the Fermi–Dirac distribution. To obtain the total rate constant, we have to integrate over all energies ϵ: Thus, the second factor in Eq. (10.13) has the meaning of a probability distribution for the transition. However, this distribution is not properly normalized to unity, and an integration would give as an extra dimension of energy. To normalize the distribution, we note that it has the form of a Gaussian, and hence the normalizing factor is $(4\pi\lambda kT)^{-1/2}$. The pre-exponential factor can be obtained from first order perturbation theory, and is $\Delta/\hbar$. Finally, we again have to convert the bulk to surface concentration, which gives an extra factor of the order of $A = 10^{-10} - 10^{-9}$ cm. So we obtain for the total rate constant:

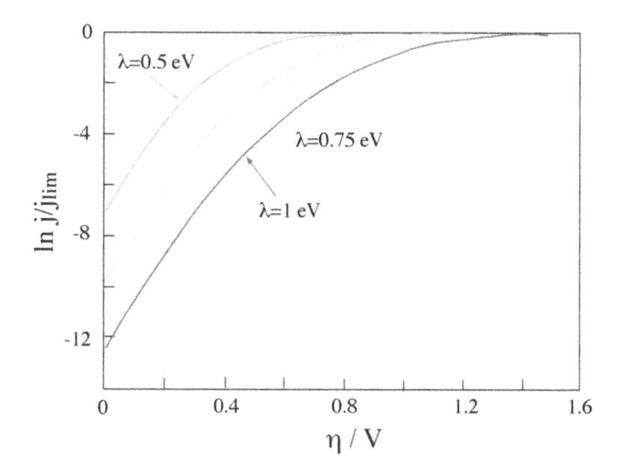

Fig. 10.6. The anodic current density as a function of electrode potential according to Eq. (10.14).

$$k_{\text{ox}} = A\frac{\Delta}{\hbar}(4\pi\lambda kT)^{-1/2}\int [1 - f(\epsilon)] \exp - \frac{(\lambda + \epsilon - e_0\eta)^2}{4\lambda kT}\, d\epsilon \qquad (10.14)$$

The integral is to be performed over the conduction band of the metal; in practice the limits can be extended to $\pm\infty$, since the integrand is negligible far from the Fermi level.

The rate constant for the reduction can be obtained from Eq. (10.10); alternatively we can note that we must change the signs of ϵ and η, and that the rate for a given ϵ must be proportional to $f(\epsilon)$. This gives:

$$k_{\text{red}} = A\frac{\Delta}{\hbar}(4\pi\lambda kT)^{-1/2}\int f(\epsilon) \exp - \frac{(\lambda - \epsilon + e_0\eta)^2}{4\lambda kT}\, d\epsilon \qquad (10.15)$$

Equations (10.14) and (10.15) are the general relations for the rate constants in the non-adiabatic case. The resulting current-potential curves for the anodic direction are shown in Fig. 10.6. They have been normalized by the constant limiting value at high overpotentials. The corresponding curves for the adiabatic case are very similar, so we do not show them separately.

There are two useful approximations: For small η, only the region near the Fermi level contributes; it is sufficient to keep first-order terms in ϵ and η in the energy of activation. The integral can then be performed explicitly, resulting in:

$$k_{\text{ox}} = A\frac{\Delta}{\hbar}\left(\frac{\pi kT}{4\lambda}\right)^{1/2} \exp\left(-\frac{\lambda - 2e_0\eta}{4kT}\right), \quad \text{for} \quad e_0\eta \ll \lambda \qquad (10.16)$$

where the integration limits have been extended to $\pm\infty$. Again, this equation has the form of the familiar Butler–Volmer law with a transfer coefficient of one-half.

A good approximation to the current-potential curve is obtained by replacing the Fermi–Dirac distribution with a step function: which results in:

$$k_{\mathrm{ox}} = A\frac{\Delta}{2\hbar}\mathrm{erfc}\frac{\lambda - e_0\eta}{(4\lambda kT)^{1/2}} \tag{10.17}$$

where

$$\begin{aligned}\mathrm{erfc}(x) &= \frac{2}{\sqrt{\pi}}\int_x^{\infty} \exp(-y^2)\, dy = 1 - \mathrm{erf}(x) \\ &= 1 - \frac{2}{\sqrt{\pi}}\int_0^{x} \exp(-y^2)\, dy\end{aligned}$$

is the compliment of the error function $\mathrm{erf}(x)$. Equation (10.17) is a good approximation in the region $e_0\eta \gg kT$. In particular we obtain at very large overpotentials a limiting rate:

$$k_{\mathrm{lim}} = A\frac{\Delta}{2\hbar}, \quad \text{for} \quad e_0\eta \gg \lambda \tag{10.18}$$

which is independent of the applied potential.

The corresponding expressions for the reduction are:

$$k_{\mathrm{red}} = A\frac{\Delta}{\hbar}\left(\frac{\pi kT}{4\lambda}\right)^{1/2}\exp\left(-\frac{\lambda+2e_0\eta}{4kT}\right), \quad \text{for} \quad |e_0\eta| \ll \lambda \tag{10.19}$$

$$k_{\mathrm{red}} = A\frac{\Delta}{2\hbar}\mathrm{erfc}\frac{e_0\eta+\lambda}{(4\lambda kT)^{1/2}}, \quad \text{for} \quad |e_0\eta| \gg kT \tag{10.20}$$

For reasons of symmetry, the limiting rates are the same in both directions.

The current-potential relation in Fig. 10.6 show Butler–Volmer behavior for small overpotentials, and limiting currents for large overpotentials, and have the same form for adiabatic and non-adiabatic reactions alike. The two kinds of reactions differ principally in the pre-exponential factors, which for an adiabatic reaction are independent of the electrode material, and for a non-adiabatic reaction depend on the strength of the interaction.

A direct comparison of the form of the current-potential curves with experiments is not easy. At low overpotentials one always observes Butler–Volmer behavior in agreement with the theory. At high overpotentials is difficult to measure kinetic currents since then the reaction is fast and usually transport controlled (see Chap. 19). The deviations from the Butler–Volmer equation predicted by theory were doubted for some time. But they have now been observed beyond doubt, and we shall review some relevant experimental results in Chap. 12, where we shall also concern ourselves with the question of adiabaticity.

The model presented here is simplified in several ways: harmonic approximation, purely classical treatment reorganization. But it does explain the basic features of electron-transfer reactions, relates the observed energies of activation to the reorganization of the inner and outer sphere, and does predict the correct form of the current-potential relationship. In some cases the energy of reorganization can be estimated (see the following), and then quantitative comparisons between theory and experiment can be made.

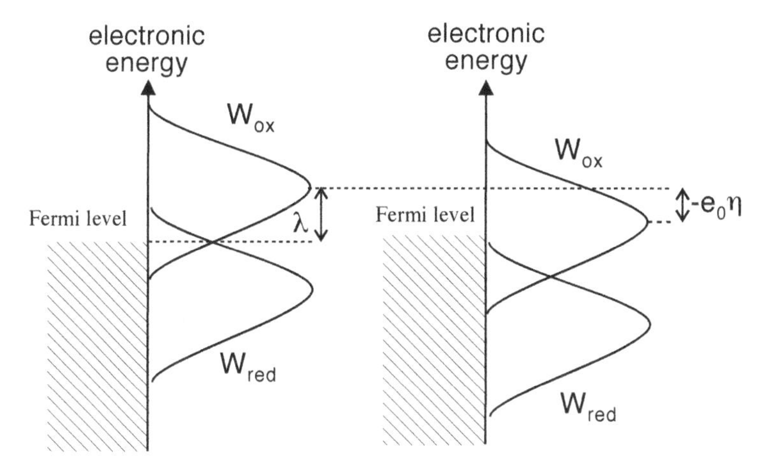

Fig. 10.7. Distributions W_{ox} and W_{red} at equilibrium (*left*) and after application of a cathodic overpotential.

10.5 Gerischer's formulation

The equations for the rate constants in the non-adiabatic case derived above have a suggestive interpretation proposed by Gerischer [4]. In the expression for the oxidation rate the term $[1 - f(\epsilon)]$ is the probability to find an empty state of energy ϵ on the electrode surface. If one interprets:

$$W_{\text{red}}(\epsilon, \eta) = (4\pi\lambda kT)^{-1/2} \exp\left(-\frac{(\lambda + \epsilon - e_0\eta)^2}{4\lambda kT}\right) \tag{10.21}$$

as the (normalized) probability of finding an occupied (reduced) state of energy ϵ in the solution, then the anodic rate is simply proportional to the probability of finding an occupied state of energy ϵ in the solution multiplied by the probability to find an empty state of energy ϵ on the metal. The maximum of W_{red} is at $\epsilon = -\lambda + e_0\eta$; so application of an overpotential shifts it by an amount $e_0\eta$ with respect to the Fermi level of the metal. *Mutatis mutandis*, the same argument can be made for the cathodic direction by defining:

$$W_{\text{ox}}(\epsilon, \eta) = (4\pi\lambda kT)^{-1/2} \exp\left(-\frac{(\lambda - \epsilon + e_0\eta)^2}{4\lambda kT}\right) \tag{10.22}$$

as the probability to find an empty (oxidized) state of energy ϵ in the solution. This has its maximum at $\epsilon = \lambda + e_0\eta$; so on application of an overpotential it is shifted by the same amount as W_{red}. Illustrations such as the one presented in Fig. 10.7 offer a useful way of visualizing simple electron-transfer reactions.

Unfortunately, the probabilities W_{ox} and W_{red} are sometimes denoted as densities of states of the oxidized and reduced species in the solution. This is a misnomer, since they have nothing to do with the electronic densities of states we have introduced earlier, and can only lead to confusion. Indeed, the two concepts are sometimes confused in the literature. Needless to say, we shall not use this terminology.

10.6 Multidimensional treatment

Here we shows that a multi-dimensional treatment of the solvent and inner sphere reorganization gives the same energy of activation. At the same time we prepare a way for estimating reorganization energies. We describe all modes that are reorganized as harmonic oscillators, which interact linearly with the charge on the reactant:

$$H_{\text{sol}} = \sum_i \left\{ \frac{1}{2}\alpha_i x_i^2 + \alpha x_i g_i (1-n) \right\} \tag{10.23}$$

For a real harmonic oscillator, we have: $\alpha_i = m_i \omega_i^2$, where m_i is the mass of the oscillator, ω_i its frequency. However, for the mathematics it is only important that the leading term of the Hamiltonian is second order, and the interaction linear. As we shall see below, the coordinate can have quite a general meaning, and for the outer sphere it is usually taken as the local polarization. As before g_i are the interaction constants with the charge on the reactant. For $n = 1$, the minimum is at $x_i = 0$, for $n = 1$ at $x_i = -g_i$. As discussed before, the mathematics can be greatly simplified by normalizing the coordinates and introducing $q_i = x_i/g_i$. Instead of Eq. (10.8) we obtain the generalized form:

$$E(q_i) = \epsilon_a n + \sum_i \{\lambda_i q_i^2 + 2\lambda_i q_i (1-n)\} \tag{10.24}$$

$\lambda_i = \alpha_i g_i^2/2$ is the contribution of the ith mode to the energy of reorganization. Equation (10.24) defines two different multi-dimensional paraboloids for $n = 1$, which is our initial state, and $n = 0$. We proceed to find the saddle point on the intersection. There, the energies on the two surfaces must be equal, which gives:

$$2\sum_i \lambda_i q_i - \epsilon_a = 0 \tag{10.25}$$

We determine the saddle point by introducing a Lagrange multiplier μ and minimize the function:

$$F(q_i) = \epsilon_a + \sum_i \lambda_i q_i^2 + \mu \left(2\sum_i \lambda_i q_i - \epsilon_a \right) \tag{10.26}$$

subject to the constraint of Eq. (10.25). This gives:

$$q_i = -\mu \qquad \mu = -\frac{\epsilon_a}{2\lambda} \qquad E_{\text{act}} = \frac{\epsilon_a^2}{4\lambda} = \frac{(-\lambda + e_0\eta)^2}{4\lambda} \tag{10.27}$$

where $\lambda = \sum_i \lambda_i$ is the total energy of reorganization. So we have obtained the same energy of activation as before. The reaction path is given by the straight line $q_i = q$, where q runs from 0 to -1 and can be identified with the generalized solvent coordinate. Remember that at equilibrium $\epsilon_a = -\lambda$; the saddle point lies between the minima for the initial and the final state only for $\epsilon_a < 0$.

10.7 The energy of reorganization

The contribution of a mode i to the energy of reorganization has been defined above as $\lambda_i = \alpha_i g_i^2/2$. During the course of the reaction, the equilibrium value of the coordinate x_i changes from $x_i = 0$ to $x_i = -g_i$. Therefore, we can also write:

$$\lambda_i = \alpha_i (\Delta q_i)^2/2 \tag{10.28}$$

where Δq_i is the change in the equilibrium position, and $\alpha_i/2$ is the coefficient of the quadratic term. This equation can be used for any type of harmonic oscillator, irrespective of the meaning of the coordinate.

Since the energy of reorganization plays a central role in electron-transfer reactions, it is useful to obtain rough estimates for specific systems. As outlined, it contains two contributions: one from the inner and one from the outer sphere. The former is readily calculated from the previous section. As an example, we consider the reaction of the $[Fe(H_2O)_6]^{2+/3+}$ couple. During the reaction the distance of the water ligands from the central ion changes; this corresponds to a reorganization of the totally symmetric, or "breathing", mode of the complex, and this seems to be the only mode which undergoes substantial reorganization. Let m be the effective mass of this mode, Δq the change in the equilibrium distance, and ω the frequency. The energy of reorganization of the inner sphere is then:

$$\lambda_{\text{in}} = \frac{1}{2} m \omega^2 (\Delta q)^2 \tag{10.29}$$

There is a small complication in that the frequency ω is different for the reduced and oxidized states; so that one has to take an average frequency. Marcus has suggested taking $\omega_{\text{av}} = 2\omega_{\text{ox}}\omega_{\text{red}}/(\omega_{\text{ox}}+\omega_{\text{red}})$. When several inner-sphere modes are reorganized, one simply sums over the various contributions. The matter becomes complicated if the complex is severely distorted during the reaction, and the two states have different normal coordinates. While the theory can be suitably modified to account for this case, the mathematics are cumbersome.

To obtain an estimate for the energy of reorganization of the outer sphere, we start from the Born model, in which the solvation of an ion is viewed as resulting from the Coulomb interaction of the ionic charge with the polarization of the solvent. This polarization contains two contributions: one is from the electronic polarizability of the solvent molecules; the other is caused by the orientation and distortion of the solvent molecules in an external field. The former is also denoted as the *fast polarization*, since it is electronic in origin and reacts on a time scale of 10^{-15}–10^{-16} s, so that it reacts practically instantly to the electron transfer; the latter is called the *slow polarization* since it is caused by the movement of atoms on a time scale of 10^{-11}–10^{-14} s. To obtain separate expressions for the two components we start with the constitutive relation between the electric field vector $\mathbf{E}$, the dielectric displacement $\mathbf{D}$, and the polarization $\mathbf{P}$:

$$\mathbf{D} = \epsilon\epsilon_0\mathbf{E} = \epsilon_0\mathbf{E} + \mathbf{P}, \quad \text{or} \quad \mathbf{P} = \left(1 - \frac{1}{\epsilon}\right)\mathbf{D} \tag{10.30}$$

where ϵ is the dielectric constant of the medium.[2] If we apply an alternating external field with a high frequency in the optical region, only the electronic polarization can follow, and the optical value ϵ_∞ of the dielectric constant applies ($\epsilon_\infty = 1.88$ for water). So the fast polarization is:

$$\mathbf{P}_f = \left(1 - \frac{1}{\epsilon_\infty}\right)\mathbf{D} \tag{10.31}$$

In a static field both components of the polarization contribute, and the static value ϵ_s of the dielectric constant must be used in Eq. (10.30). The slow polarization is obtained by subtracting $\mathbf{P}_f$, which gives:

$$\mathbf{P}_s = \left(\frac{1}{\epsilon_\infty} - \frac{1}{\epsilon_s}\right)\mathbf{D} \tag{10.32}$$

The reorganization of the solvent molecules can be expressed through the change in the slow polarization. Consider a small volume element ΔV of the solvent in the vicinity of the reactant; it has a dipole moment $\mathbf{m} = \mathbf{P}_s\,\Delta V$ caused by the slow polarization, and its energy of interaction with the external field $\mathbf{E}_{\text{ex}}$ caused by the reacting ion is $-\mathbf{P}_s \cdot \mathbf{E}_{\text{ex}}\,\Delta V = -\mathbf{P}_s \cdot \mathbf{D}\,\Delta V/\epsilon_0$, since $\mathbf{E}_{\text{ex}} = \mathbf{D}/\epsilon_0$. We take the polarization $\mathbf{P}_s$ as the relevant outer-sphere coordinate, and require an expression for the contribution ΔU of the volume element to the potential energy of the system. In the harmonic approximation this must be a second-order polynomial in $\mathbf{P}_s$, and the linear term is the interaction with the external field, so that the equilibrium values of $\mathbf{P}_s$ in the absence of a field vanishes:

$$\Delta U/\Delta V = \frac{1}{2}\alpha\mathbf{P}_s^2 - \mathbf{P}_s \cdot \mathbf{D}/\epsilon_0 + C \tag{10.33}$$

where C is independent of $\mathbf{P}_s$, and the constant α is still to be determined. For this purpose we calculate the equilibrium value of the slow polarization by minimizing ΔU and identifying this result with the value from Eq. (10.32):

[2] We use the usual symbol ϵ for the dielectric constant; no confusion should arise with the energy variable employed

$$\mathbf{P}_s^{\mathrm{eq}} = \frac{\mathbf{D}}{\alpha\epsilon_0}, \quad \text{hence} \quad \frac{1}{\alpha\epsilon_0} = \left(\frac{1}{\epsilon_\infty} - \frac{1}{\epsilon_s}\right) \tag{10.34}$$

During the reaction the dielectric displacement changes from $\mathbf{D}_{\mathrm{ox}}$ to $\mathbf{D}_{\mathrm{red}}$ (or vice versa), and the equilibrium value from $\mathbf{D}_{\mathrm{ox}}/\alpha\epsilon_0$ to $\mathbf{D}_{\mathrm{red}}/\alpha\epsilon_0$. Therefore the contribution of the volume element ΔV to the energy of reorganization of the outer sphere is:

$$\Delta\lambda_{\mathrm{out}} = \frac{1}{2\epsilon_0}\left(\frac{1}{\epsilon_\infty} - \frac{1}{\epsilon_s}\right)\left(\mathbf{D}_{\mathrm{ox}} - \mathbf{D}_{\mathrm{red}}\right)^2 \Delta V \tag{10.35}$$

The total energy of reorganization of the outer sphere is obtained by integrating over the volume of the solution surrounding the reactant:

$$\lambda_{\mathrm{out}} = \frac{1}{2\epsilon_0}\left(\frac{1}{\epsilon_\infty} - \frac{1}{\epsilon_s}\right)\int \left(\mathbf{D}_{\mathrm{ox}} - \mathbf{D}_{\mathrm{red}}\right)^2 dV \tag{10.36}$$

The dielectric displacement must be calculated from electrostatics; for a reactant in front of a metal surface the image force has to be considered. For the simple case of a spherical ion in front of a metal electrode experiencing the full image interaction, a straightforward calculation gives:

$$\lambda_{\mathrm{out}} = \frac{e_0^2}{8\pi\epsilon_0}\left(\frac{1}{\epsilon_\infty} - \frac{1}{\epsilon_s}\right)\left(\frac{1}{a} - \frac{1}{2d}\right) \tag{10.37}$$

where a is the radius of the ion, and d the distance from the metal surface. Because of the use of macroscopic electrostatics, this equation should be viewed as providing no more than an estimate for λ_{out}.

10.8 Adiabatic versus non-adiabatic transitions

If a reaction proceeds adiabatically or not, depends on the strength Δ of the electronic interaction between the reactant and the electrode surface, which also determines the width of the density of states of the reacting level. For small interactions, first order perturbation theory holds, and the pre-exponential factor is proportional to Δ – see Eq. (10.14). For large Δ, the pre-exponential factor is determined by solvent dynamics. The latter can be described in terms of Kramer's theory, which we cannot treat here in any detail. Briefly, an important factor is the solvent *friction*, which determines the typical time that the solvent takes to reorient. The higher the friction, the lower is the pre-exponential factor A. As mentioned before, for aqueous solutions a value of $A \approx 10^3$ cm s^{-1} seems to be a good estimate.

The relation between the rate constant k and the interaction strength Δ is shown in Fig. 10.8, which is based on a computer simulation [5]. For low interactions, the rate follows perturbation theory. Then, solvent dynamics starts to influence the rate, and in a certain region it is independent of Δ. The

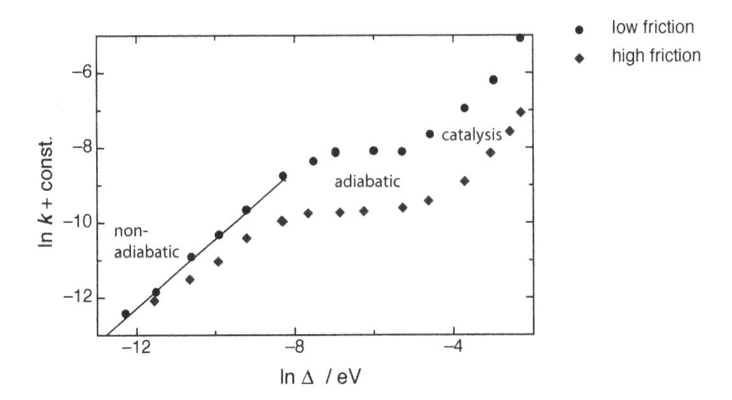

Fig. 10.8. Dependence of the rate constant on the interaction strength Δ. The *black line* shows the prediction from 1st order perturbation theory.

height of this plateau region depends on the friction: the higher the friction, the lower the plateau. This is the region in which the original theories of Marcus and Hush hold. For very high interactions, the last term of Eq. (10.8) comes into play and lowers the energy of activation, so the rate rises again. The latter is a catalytic effect, caused by the electronic interaction with the metal.

On bare metal surfaces, outer sphere reactions are typically adiabatic and fall into the Marcus–Hush plateau region. Since they are independent of Δ, they also do not depend on the nature of the metal. On semiconductors, semi-metals like graphite, and in particularly on electrodes covered by an insulating film, they can proceed non-adiabatically. The catalytic region does not seem to play a role for outer-sphere reactions – they are so fast that they do not need to be catalysed. However, catalysis is very important for inner sphere reactions like hydrogen evolution, which we will consider in detail later.

Problems

1. Consider a one-dimensional system in which the potential energy functions for the oxidized and reduced states are:

$$U_{\text{ox}}(q) = e_{\text{ox}} + \frac{1}{2}m\omega^2 q^2$$
$$U_{\text{red}}(q) = e_{\text{red}} + \frac{1}{2}m\omega^2 (q-\delta)^2$$

Calculate the intersection point of these two parabolas and define the energy of reorganization. Calculate the energies of activation for the forward and the backward direction.

2. Assume that the current-potential curves of a system are given by Eqs. (10.17) and (10.20). Calculate the effective transfer coefficients defined by:

$$\alpha = \frac{kT}{e_0} \frac{\partial \ln j_a}{\partial \eta} \qquad \beta = -\frac{kT}{e_0} \frac{\partial \ln |j_c|}{\partial \eta}$$

Their values depend on the overpotential. Show that for $\eta = 0$: $\alpha + \beta \neq 1$. This (small) error arises because the Fermi–Dirac distribution has been replaced by a step function.

3. From Eq. (10.36) calculate the energy of reorganization of a single spherical reactant in the bulk of a solution. Derive Eq. (10.37) for a reactant in front of a metal electrode.
4. Show that for $\epsilon_\infty = 1$ Eq. (10.36) reduces to the Born equation for the energy of solvation of an ion.

References

1. R.A. Marcus, *J. Chem. Phys.* **24** (1956) 966.
2. N.S. Hush, *J. Chem. Phys.* **28** (1958) 962.
3. W. Schmickler, *J. Electroanal. Chem.* **204** (1986) 31.
4. H. Gerischer, *Z. Phys. Chem.* NF **26** (1960) 223; **27** (1961) 40, 48.
5. W. Schmickler and J. Mohr, *J. Chem. Phys.* **117** (2002) 2867.

11

The semiconductor-electrolyte interface

11.1 Electrochemistry at semiconductors

Many naturally occurring substances, in particular the oxide films that form spontaneously on some metals, are semiconductors. Also, electrochemical reactions are used in the production of semiconductor chips, and recently semiconductors have been used in the construction of electrochemical photocells. So there are good technological reasons to study the interface between a semiconductor and an electrolyte. Our main interest, however, lies in more fundamental questions: How does the electronic structure of the electrode influence the properties of the electrochemical interface, and how does it affect electrochemical reactions? What new processes can occur at semiconductors that are not known from metals?

11.2 Potential profile and band bending

When a semiconducting electrode is brought into contact with an electrolyte solution, a potential difference is established at the interface. The conductivity even of doped semiconductors is usually well below that of an electrolyte solution; so practically all of the potential drop occurs in the boundary layer of the electrode, and very little on the solution side of the interface (see Fig. 11.1). The situation is opposite to that on metal electrodes, but very similar to that at the interface between a semiconductor and a metal.

The variation of the electrostatic potential $\phi(x)$ in the surface region entails a bending of the bands, since the potential contributes a term $-e_0\phi(x)$ to the electronic energy. Consider the case of an n-type semiconductor. We set $\phi = 0$ in the bulk of the semiconductor. If the value ϕ_s of the potential at the surface is positive, the bands band downwards, and the concentration of electrons in the conduction band is enhanced (see Fig. 11.2). This is called an *enrichment layer*. If $\phi_s < 0$, the bands bend upward, and the concentration of electrons at the surface is reduced; we speak of a *depletion layer*. On the other

W. Schmickler, E. Santos, *Interfacial Electrochemistry*, 2nd ed.,
DOI 10.1007/978-3-642-04937-8_11, © Springer-Verlag Berlin Heidelberg 2010

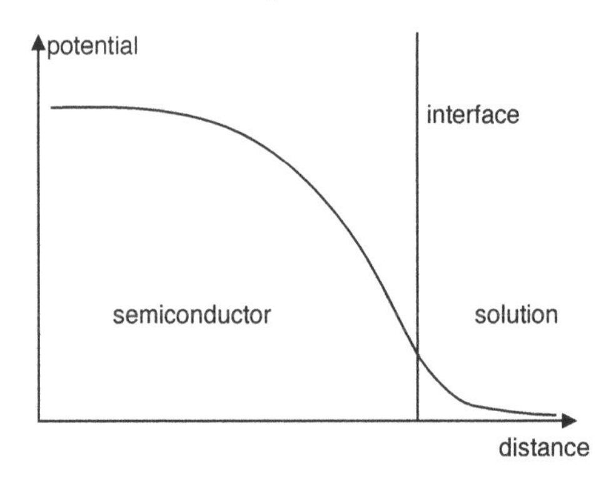

Fig. 11.1. Variation of the potential at the semiconductor-solution interface (schematic).

hand, the concentration of the holes, the minority carriers, is enhanced at the surface; if it exceeds that of the electrons, one speaks of an *inversion layer*. The special potential at which the electrostatic potential is constant (i.e., $\phi(x) = 0$ throughout the semiconductor), is the *flat-band potential*, which is equivalent to the potential of zero charge. In Chap. 4 we noted that, because of the occurrence of dipole potentials, the difference in outer potential does not vanish at the pzc; the same is true for the flat-band potential of a semiconductor in contact with an electrolyte solution.

Mutatis mutandis the same terminology is applied to the surface of p-type semiconductors. So if the bands bend upward, we speak of an enrichment layer; if they bend downward, of a depletion layer.

Just as in Gouy–Chapman theory, the variation of the potential can be calculated from Poisson's equation and Boltzmann statistics (in the nondegenerate case). As an example we consider an n-type semiconductor, and limit ourselves to the case where the donors are completely ionized, and the concentration of holes is negligible throughout – a full treatment of all possible cases is given in [1, 2]. The charge density in the space-charge region is the sum of the static positive charge on the ionized donors, and the mobile negative charge of the conduction electrons. Let n_b be the density of electrons in the bulk, which equals the density of donors since the bulk is electroneutral. Poisson's equation gives:

$$\frac{d^2\phi}{dx^2} = -\frac{n_{\mathrm{b}}}{\epsilon\epsilon_0}\left(1 - \exp\frac{e_0\phi}{kT}\right) \tag{11.1}$$

which is reminiscent of the Poisson–Boltzmann equation. An approximate analytic solution can be derived for a depletion layer; the band has a parabolic shape, and the corresponding interfacial capacitance C_{sc} is given by the *Mott–Schottky* equation (see Appendix), which is usually written in the form:

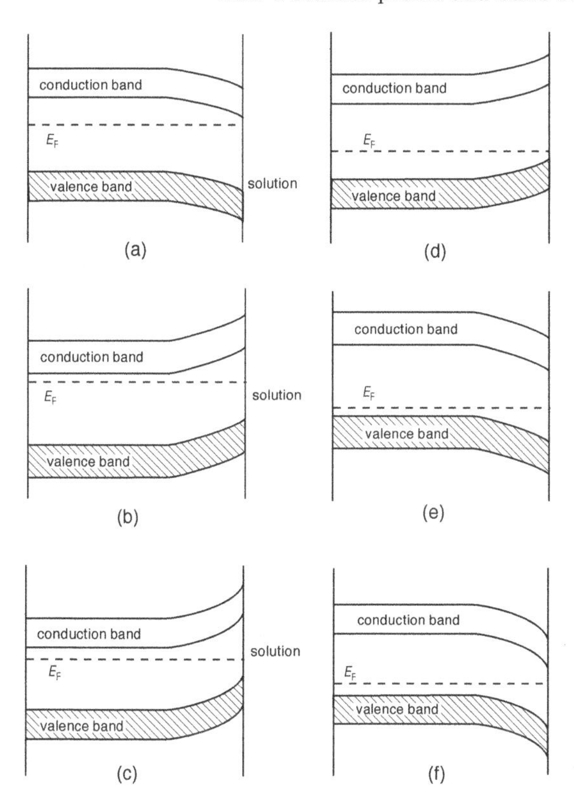

Fig. 11.2. Band bending at the interface between a semiconductor and an electrolyte solution; (**a**)–(**c**) n-type semiconductor: (**a**) enrichment layer, (**b**) depletion layer, (**c**) inversion layer; (**d**)–(**f**) p-type semiconductor: (**d**) enrichment layer, (**e**) depletion layer, (**f**) inversion layer.

$$\left(\frac{1}{C_{sc}}\right)^2 = \frac{2}{\epsilon\epsilon_0 e_0 n_b}\left(|\phi_s| - \frac{kT}{e_0}\right) \tag{11.2}$$

Often, the small term kT/e_0 is neglected. The total interfacial capacity C is a series combination of the space-charge capacities C_{sc} of the semiconductor and C_{sol} of the solution side of the interface. However, generally $C_{sol} \gg C_{sc}$, and the contribution of the solution can be neglected. Then a plot of $1/C^2$ versus the electrode potential ϕ (which differs from ϕ_s by a constant) gives a straight line (see Fig. 11.3). From the intercept with the ϕ axis the flat-band potential is determined; if the dielectric constant ϵ is known, the donor density can be calculated from the slope. The same relation holds for the depletion layer of a p-type semiconductor.

Semiconductors that are used in electrochemical systems often do not meet the ideal conditions on which the Mott–Schottky equation is based. This is particularly true if the semiconductor is an oxide film formed in situ by oxidiz-

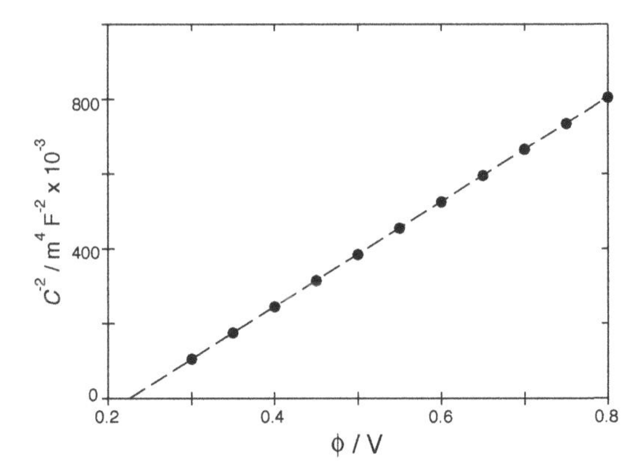

Fig. 11.3. Mott–Schottky plot for the depletion layer of an n-type semiconductor; the flat-band potential E_{fb} is at 0.2 V. The data extrapolate to $E_{fb} + kT/e_0$.

ing a metal such as Fe or Ti. Such semiconducting films are often amorphous, and contain localized states in the band gap that are spread over a whole range of energies. This may give rise to a frequency dependence of the space-charge capacity, because localized states with low energies have longer time constants for charging and discharging. It is therefore important to check that the interfacial capacity is independent of the frequency if one wants to determine donor densities from Eq. (11.2).

11.3 Electron-transfer reactions

There is a fundamental difference between electron-transfer reactions on metals and on semiconductors. On metals the variation of the electrode potential causes a corresponding change in the molar Gibbs energy of the reaction. Due to the comparatively low conductivity of semiconductors, the positions of the band edges at the semiconductor surface do not change with respect to the solution as the potential is varied. However, the relative position of the Fermi level in the semiconductor is changed, and so are the densities of electrons and holes on the semiconductor surface.

The general shape of the current-potential curves for a perfect, non-degenerate semiconductor, for which the Fermi level lies well within the band gap, is easily derived. We first consider electron exchange with the conduction band. Since concentration of electrons in this band is very low, electron transfer from a redox couple in the solution to this band is not impeded by them. Further, since the relative position of the electronic levels in the solution and the semiconductor surface do not change with potential, the anodic current is constant, and we call its density j_0^c, the superscript indicating the conduction band. On the other, application of a negative overpotential η brings the band edge at the surface by an amount $e_0\eta$ closer to the Fermi level, and the concentration of electrons increases exponentially. Noting that for $\eta = 0$ the total current must vanish, we can write the current density passing through the conduction band as:

$$j^c = j_0^c \left[1 - \exp\left(-\frac{e_0\eta}{kT}\right)\right] \tag{11.3}$$

Obviously, this current-potential characteristics has rectifying properties (see Fig. 11.5).

Conversely, the valence band is practically full, and electron transfer from this band to the solution is constant; the corresponding current density we call $-j_0^v$. Electron transfer from the solution to the valence band is proportional to the density of holes in this band, which increases exponentially with $e_0\eta$. Therefore we obtain for the current through the valence band:

$$j^v = j_0^v \left[\exp\frac{e_0\eta}{kT} - 1\right] \tag{11.4}$$

Gerischer's terminology is popular in semiconductor electrochemistry, and it is instructive to calculate the currents in this model. This implies that the transfer is non-adiabatic, which seems plausible in view of the fact that the surface orbitals of semiconductors are less extended than those of metals.

We start from Eq. (10.14) for the rate of electron transfer from a reduced state in the solution to a state of energy ϵ on the electrode, and rewrite it in the form:

$$k_{\text{ox}}(\epsilon) = A' \int [1 - f(\epsilon)]\, W_{\text{red}}(\epsilon, \eta) d\epsilon \tag{11.5}$$

using Gerischer's terminology; Fig. 11.4 shows a corresponding plot. We have introduced $A' = A\Delta/\hbar$ for brevity. We still have to specify the integration limits. There are two contributions to the anodic current density, j_a^v from the valence and j_a^c from the conduction band. Denoting by E_v, E_c the band edges at the surface, we write for the current density:

$$j_a^v = FA'c \int_{-\infty}^{E_v - E_F} d\epsilon\ [1 - f(\epsilon)]\, W_{\text{red}}(\epsilon, \eta) \tag{11.6}$$

$$j_a^c = FAc \int_{E_c - E_F}^{\infty} d\epsilon\ [1 - f(\epsilon)]\, W_{\text{red}}(\epsilon, \eta) \tag{11.7}$$

Strictly speaking, the integrals should extend over the two bands only; however, far from the band edges the integrands are small; so the integration regions may safely be extended to infinity. The band edges E_v and E_c are measured with respect to the Fermi level of the electrode, and move with the overpotential; they are fixed with respect to the Fermi level of the redox couple in the solution. Writing $\Delta E_v = E_F - E_v(\eta = 0)$ and $\Delta E_c = E_c(\eta = 0) - E_F$, we have: $E_v - E_F = -\Delta E_v + e_0\eta$, $E_c - E_F = \Delta E_c + e_0\eta$. In the valence band $[1 - f(\epsilon)] \approx \exp[\epsilon/kT]$, in the conduction band $[1 - f(\epsilon)] \approx 1$, both approximations hold for nondegenerate semiconductors only. This gives:

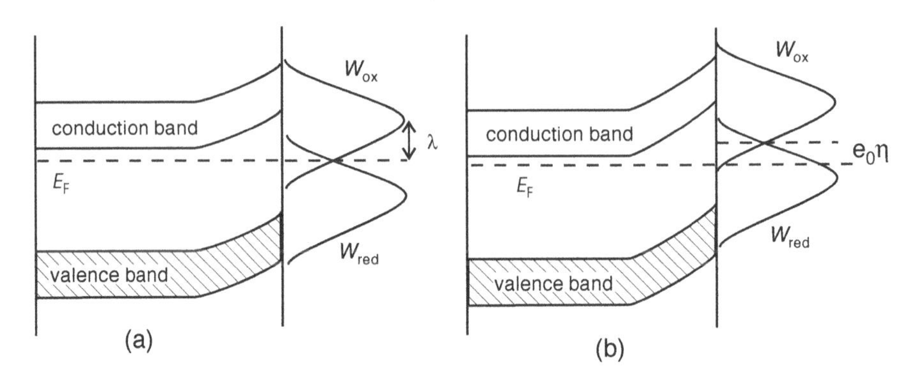

Fig. 11.4. Gerischer diagram for a redox reaction at an n-type semiconductor: (**a**) at equilibrium the Fermi levels of the semiconductor and of the redox couple are equal; (**b**) after application of an anodic overpotential.

$$j_a^v = FA'c \int_{-\infty}^{-\Delta E_v + e_0\eta} d\epsilon \, \exp\frac{\epsilon}{kT} \, W_{\text{red}}(\epsilon, \eta) \tag{11.8}$$

$$j_a^c = FA'c \int_{\Delta E_c + e_0\eta}^{\infty} d\epsilon \, W_{\text{red}}(\epsilon, \eta) \tag{11.9}$$

We substitute $\xi = \epsilon - e_0\eta$, and note that $W_{\text{red}}(\epsilon, \eta) = W_{\text{red}}(\epsilon - e_0\eta, 0)$:

$$\begin{aligned} j_a^v(\eta) &= FA'c \int_{-\infty}^{-\Delta E_v} d\xi \exp\frac{\xi + e_0\eta}{kT} W_{\text{red}}(\xi, 0) \\ &= j_a^v(\eta = 0) \exp\frac{e_0\eta}{kT} \end{aligned} \tag{11.10}$$

$$j_a^c(\eta) = FA'c \int_{\Delta E_c}^{\infty} d\xi \; W_{\text{red}}(\xi, 0) = j_a^c(\eta = 0) \tag{11.11}$$

So, as already discussed above, the contribution of the valence band to the anodic current increases exponentially with the applied potential, because the number of holes that can accept electrons increases. In contrast, the anodic current via the conduction band is unchanged, since it remains practically empty. These equations hold independent of the particular form of the function W_{red}. Similarly the contributions of the valence and conduction bands to the cathodic current densities are:

$$\begin{aligned} j_c^v(\eta) &= FA'c \int_{-\infty}^{-\Delta E_v} d\xi \; W_{\text{ox}}(\xi, 0) \\ &= j_c^v(\eta = 0) \end{aligned} \tag{11.12}$$

$$\begin{aligned} j_c^c(\eta) &= FA'c \int_{\Delta E_c}^{\infty} d\xi \exp\left(-\frac{\xi + e_0\eta}{kT}\right) \; W_{\text{ox}}(\xi, 0) \\ &= j_c^c(\eta = 0) \exp\left(-\frac{e_0\eta}{kT}\right) \end{aligned} \tag{11.13}$$

The contribution of the valence band does not change when the overpotential is varied, since it remains practically completely filled. In contrast, the contribution of the conduction band decreases exponentially with η (or increases exponentially with $-\eta$) because of the corresponding change of the density of electrons (Fig. 11.5). All equations derived in this section hold only as long as the surface is nondegenerate; that is, the Fermi level does not come close to one of the bands.

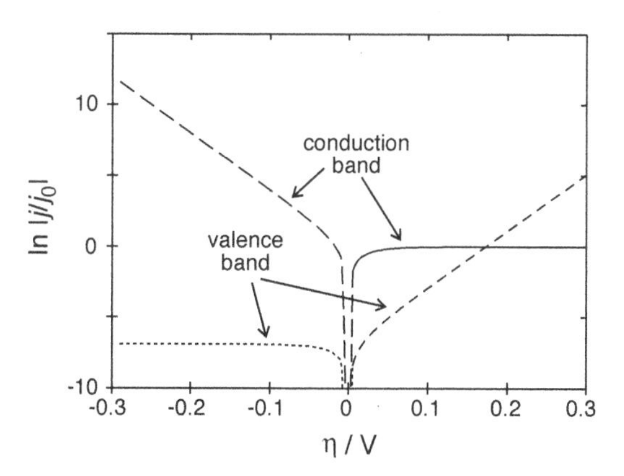

Fig. 11.5. Current-potential characteristics for a redox reaction via the conduction band or via the valence band. The current was normalized by setting $j_0^v = 1$. In this example the redox system overlaps more strongly with the conduction than with the valence band.

Typically the contributions of the two bands to the current are of rather unequal magnitude, and one of them dominates the current. Unless the electronic densities of states of the two bands differ greatly, the major part of the current will come from the band that is closer to the Fermi level of the redox system (see Fig. 11.4). The relative magnitudes of the current densities at vanishing overpotential can be estimated from the explicit expressions for the distribution functions W_{red} and W_{ox}:

$$
\begin{aligned}
j_0^v &= FA'c \int_{-\infty}^{-\Delta E_v} d\xi \; W_{\mathrm{ox}}(\xi, 0) \\
&= 2FA'c \operatorname{erfc} \frac{\lambda + \Delta \mathrm{E_v}}{(4\lambda \mathrm{kT})^{1/2}}
\end{aligned}
\tag{11.14}
$$

$$
\begin{aligned}
j_0^c &= FA'c \int_{\Delta E_c}^{\infty} d\xi \; W_{\mathrm{red}}(\xi, 0) \\
&= 2FA\rho_c \operatorname{erfc} \frac{\lambda + \Delta \mathrm{E_c}}{(4\lambda \mathrm{kT})^{1/2}}
\end{aligned}
\tag{11.15}
$$

If the electronic properties of the semiconductor – the Fermi level, the positions of the valence and the conduction band, and the flat-band potential – and those of the redox couple – Fermi level and energy of reorganization – are known, the Gerischer [3] diagram can be constructed, and the overlap of the two distribution functions W_{ox} and W_{red} with the bands can be calculated.

Both contributions to the current obey the Butler–Volmer law. The current flowing through the conduction band has a vanishing anodic transfer coefficient, $\alpha_c = 0$, and a cathodic coefficient of unity, $\beta_c = 1$. Conversely, the current through the valence band has $\alpha_v = 1$ and $\beta_v = 0$. Real systems do not always show this perfect behavior. There can be various reasons for this; we list a few of the more common ones:

1. Electronic surface states may exist at the interface; they give rise to an additional capacity, so that the band edges at the surface change their energies with respect to the solution.
2. When the semiconductor is highly doped, the space-charge region is thin, and electrons can tunnel through the barrier formed at a depletion layer.
3. At high current densities the transport of electrons and holes may be too slow to establish electronic equilibrium at the semiconductor surface.
4. The semiconductor may be amorphous, in which case there are no sharp band edges.

An example of an electron-transfer reaction on a semiconductor electrode will be given in the next chapter.

11.4 Photoinduced electron transfer

Semiconducting electrodes offer the intriguing possibility to enhance the rate of an electron-transfer reaction by photoexcitation. There are actually two different effects: Either charge carriers in the electrode or the redox couple can be excited. We give examples for both mechanisms.

11.4.1 Photoexcitation of the electrode

If light of a frequency ν, with $h\nu \geq E_g$, is incident on a semiconducting electrode, it can excite an electron from the valence into the conduction band, so that an electron-hole pair is created. In the space-charge region the pair can be separated by the electric field, which prevents recombination. The electrical field produces a force F in the x direction perpendicular to the surface, and the equation of motion for an electron is given by:

$$F = -e_0 E_x = \hbar \frac{dk}{dt} \tag{11.16}$$

where k is the wavevector of the electron, and $\hbar k$ its momentum. For a hole, the force is in the opposite direction.

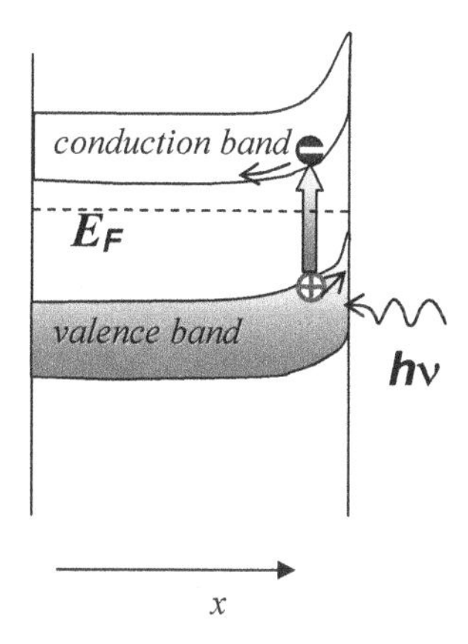

Fig. 11.6. Photogeneration of holes at the depletion layer of an n-type semiconductor.

Depending on the direction of the field, one of the carriers will migrate toward the bulk of the semiconductor, and the other will drift to the surface, where it can react with a suitable redox partner. These concepts are illustrated in Fig. 11.6 for a depletion layer of a n-type semiconductor. Holes generated in the space-charge region drift towards the surface, where they can accept electrons from a reduced species with suitable energy. According to the momentum balance for the system consisting of the electron–hole pair and the absorbed photon, we have:

$$k_e + k_h = k_{\text{ph}} \approx 0 \tag{11.17}$$

since the wavevector for a phonon with an energy of the order of a few electron volt is negligible. Thus $k_h = -k_e$, and in a band-structure plot $E(k)$ the transition is vertical. This is the typical case when the maximum of the valence band and the minimum of the conduction band coincide, and one speaks of a *direct transition* – see Fig. 11.7. The threshold for direct transitions is given by $\hbar\nu = E_g$.

When the maximum of the valence band and the minimum of the conduction band do not lie at the same wavevector k, *indirect transitions* involving a phonon may occur. The principle is depicted on the right hand side of Fig. 11.7. A phonon is needed to conserve the total momentum. The adsorption threshold for indirect transitions between the band edges is:

$$h\nu = E_g + \hbar\Omega \tag{11.18}$$

where the last term accounts for the energy of the participating phonon.

The absorption coefficient α near the band edge depends on the photon energy according to:

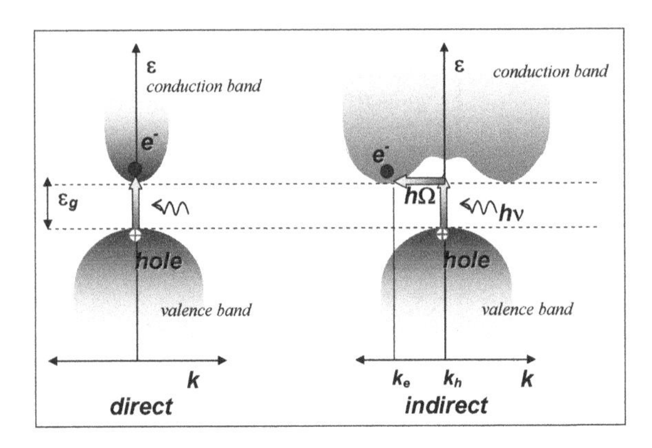

Fig. 11.7. Direct and indirect transitions

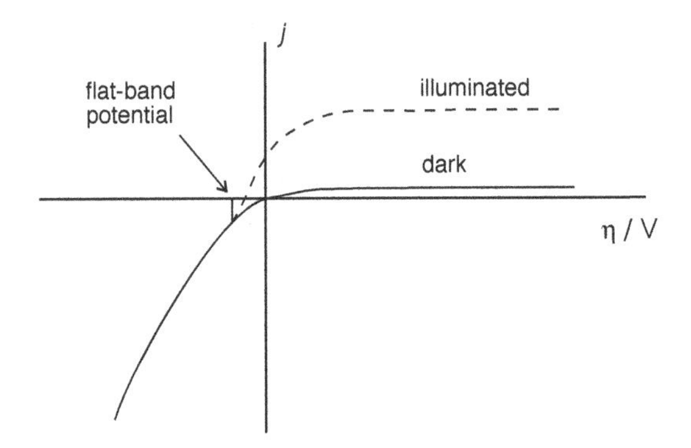

Fig. 11.8. Current-potential characteristics for an n-type semiconductor in the *dark* and *under illumination.* The difference between the two curves is the photocurrent.

$$\alpha = A\,\frac{(h\nu - E_g)^{n/2}}{h\nu} \tag{11.19}$$

We will not give the details of the derivation of this equation, which is complicated and depends on selection rules and the band structure. A is a constant and n depends on whether the transition is direct ($n = 1$) or indirect ($n = 4$).

The potential dependence of this photocurrent is shown in Fig. 11.8. It sets in at the flat-band potential and continues to rise until the band bending is so large that all the holes generated by the incident light reach the electrode surface, where they react with a suitable partner. If the reaction with the redox system is sufficiently fast, the generation of charge carriers is the rate-determining step, and the current is constant in this region.

In a real system the photocurrent can depend on a number of effects:

1. The generation of the carriers in the semiconductor.

2. The migration of the carriers in the space-charge region.
3. Diffusion of carriers that are generated outside the space-charge region.
4. Loss of carriers either by electron-hole recombination or by trapping at localized states in the band gap or at the surface.
5. The rate of the electrochemical reaction that consumes the carriers.

When all these factors contribute, the situation becomes almost hopelessly complicated. The simplest realistic case is that in which the photocarriers are generated in the space-charge region and migrate to the surface, where they are immediately consumed by an electrochemical reaction. We consider this case in greater detail. Suppose that light of frequency ν, with $h\nu > E_g$, is incident on a semiconducting electrode with unit surface area under depletion conditions (see Fig. 11.6). Let I_0 be the incident photon flux, and α the absorption coefficient of the semiconductor at frequency ν. At a distance x from the surface, the photon flux has decreased to $I_0\ \exp(-\alpha x)$, of which a fraction α is absorbed. So the rate of carrier generation is:

$$g(x) = I_0 \alpha\ \exp(-\alpha x) \tag{11.20}$$

This equation presumes that each photon absorbed creates an electron-hole pair; if there are other absorption mechanisms, the right-hand side must be multiplied by a quantum efficiency. The total rate of minority carrier generation is obtained by integrating over the space-charge region:

$$\int_0^{L_{\rm sc}} I_0 \alpha \exp\left(-\alpha x\right)\ dx = I_0 \left[1 - \exp\left(-\alpha L_{\rm sc}\right)\right] \tag{11.21}$$

where the width L_{sc} of the space charge region is (see appendix):

$$L_{\rm sc} = L_0 (\phi - \phi_{\rm fb})^{1/2}, \quad \text{with} \quad L_0 = \left(\frac{\epsilon \epsilon_0}{e_0 n_{\rm b}}\right)^{1/2} \tag{11.22}$$

so that the the photocurrent generated in the space-charge layer is:

$$j_p = e_0 I_0 \left(1 - \exp\left[-\alpha L_0 (\phi - \phi_{\rm fb})^{1/2}\right]\right) \tag{11.23}$$

In the general case there may also be a contribution due to the diffusion of carriers from the bulk. This is treated in Problem 12.3, where the concept of a *diffusion length* L_d of the minority carriers is introduced. The sum of both contribution results in:

$$j_t = e_0 I_0 \left(1 - \frac{\exp\left[-\alpha L_0 (\phi - \phi_{\rm fb})^{1/2}\right]}{1 + \alpha L_d}\right) \tag{11.24}$$

For $\alpha L_d \ll 1$ the contribution from the bulk can be neglected. If in addition $\alpha L_{\rm sc} \ll 1$ the exponential can be expanded, and the flat-band potential can be determined by plotting the square of the photocurrent versus the potential:

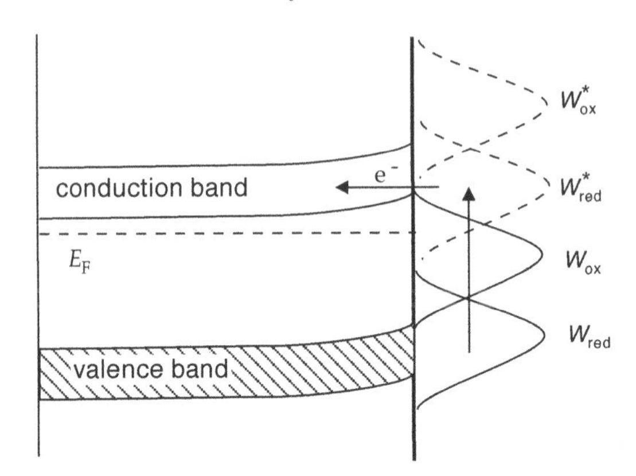

Fig. 11.9. Photoexcitation of a redox couple.

$$j_p^2 = (e_0 I_0 \alpha L_0)^2 (\phi - \phi_{fb}) \tag{11.25}$$

A plot of j_p^2 versus potential should result in a straight line, whose slope depends on the photon energy. The flat-band potential can be obtained from the intercept. We shall consider an example in Chap. 12.

11.4.2 Photoexcitation of a redox species

Another kind of photoeffect occurs if a redox system in its ground state overlaps weakly with the bands of the electrode but has an excited state which overlaps well. As an example, we consider an n-type semiconducting electrode with a depletion layer at the surface, and a reduced species red whose distribution function $W_{red}(\epsilon, \eta)$ lies well below the conduction band (see Fig. 11.9), so that the rate of electron transfer to the conduction band is low. On photoexcitation the excited state red* is produced, whose distribution function $W^*_{red}(\epsilon, \eta)$ overlaps well with the conduction band, so that it can inject electrons into this band. The electric field in the space-charge region pulls the electron into the bulk of the electrode, thus preventing recombination with the oxidized species, and a photocurrent is observed.

11.5 Dissolution of semiconductors

From a chemical point of view a hole at the surface of a semiconductor entails a missing electron and hence a partially broken bond. Consequently semiconductors tend to dissolve when holes accumulate at the surface. In particular this is true for enrichment layers of p-type material. At the depletion layers of n-type materials the holes required for the dissolution can also be produced by photoexcitation.

Such dissolution reactions usually contain several steps and are complicated. An important example is silicon. In aqueous solutions this is generally covered by an oxide film that inhibits currents and hence corrosion. However, in HF solutions it remains oxide free, and p-type silicon dissolves readily under accumulation conditions. This reaction involves two holes and two protons, the final product is Si(IV), but the details are not understood. A simpler example is the photodissolution of n-type CdS, which follows the overall reaction:

$$\mathrm{CdS} + 2\mathrm{h}^+ \rightarrow \mathrm{Cd}^{2+} + \mathrm{S} \tag{11.26}$$

under depletion conditions.

On polar semiconductors the dissolution may also involve electrons from the conduction band, leading to the production of soluble anions. For example, under accumulation conditions the dissolution of n-type CdS takes place according to the reaction scheme:

$$\mathrm{CdS} + 2\mathrm{e}^- \rightarrow \mathrm{Cd} + \mathrm{S}^{2-} \tag{11.27}$$

The dissolution of semiconductors is usually an undesirable process since it diminishes the stability of the electrode and limits their use in devices such as electrochemical photocells. On the other hand, the etching of silicon in HF solutions is a technologically important process.

Appendix: the Mott–Schottky capacity

We consider the depletion layer of an n-type semiconductor, assuming that the concentration of holes is negligible throughout. The situation is depicted in Fig. 11.10, which also defines the coordinate system employed. Starting from Eq. (11.1):

$$\frac{d^2\phi}{dx^2} = -\frac{e_0 n_\mathrm{b}}{\epsilon\epsilon_0}\left(1 - \exp\frac{e_0\phi}{kT}\right) \tag{11.28}$$

we again multiply both sides by $2d\phi/dx$, and integrate from zero to infinity, and obtain:

$$-E(0)^2 = \frac{2e_0 n_\mathrm{b}}{\epsilon\epsilon_0}\left(\phi_\mathrm{s} + \frac{kT}{e_0}\right) \tag{11.29}$$

where $\phi_s = \phi_\mathrm{s}$, and a term of the order $\exp[e_0\phi_s/kT]$ has been neglected. Noting that the potential $\phi(x)$ is negative throughout the space-charge region, we obtain:

$$\frac{\sigma}{\epsilon\epsilon_0} = \sqrt{\frac{2e_0 n_\mathrm{b}}{\epsilon\epsilon_0}}\sqrt{|\phi_\mathrm{s}| - \frac{kT}{e_0}} \tag{11.30}$$

Differentiation gives:

$$C = \frac{dq}{d\phi_\mathrm{s}} = \left(\frac{e_0 n_\mathrm{b}\epsilon\epsilon_0}{2[|\phi_\mathrm{s}| - kT/e_0]}\right)^{1/2} \tag{11.31}$$

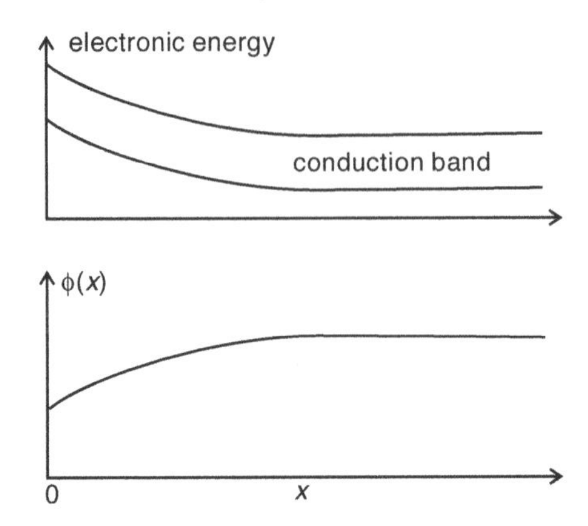

Fig. 11.10. Depletion layer at the surface of an n-type semiconductor; the surface is at $x = 0$.

which on rearranging gives Eq. (11.2).

The total width of the space-charge region can be estimated from the following consideration. Throughout the major part of the depletion region we have: $-e_0\phi \gg kT$, and the concentration of the electrons is negligible. In this region the exponential term on the right-hand side of Eq. (11.28) can be neglected, and the space charge is determined by the concentration of the donors – each donor carries a positive charge since it has given one electron to the conduction band. The band has a parabolic shape, but only the left half of the parabola has a physical meaning. The potential can be written in the form:

$$\phi(x) = -\frac{e_0 n_{\mathrm{b}}}{2\epsilon\epsilon_0} x^2 + ax + \phi_s \tag{11.32}$$

where:

$$a = \left.\frac{\partial \phi}{\partial x}\right|_{x=0} = -E(0) \tag{11.33}$$

The width L_{sc} of the space charge region is given by the position where the potential is minimal. Differentiation gives:

$$L_{\mathrm{sc}} = -\frac{\epsilon\epsilon_0}{e_0 n_{\mathrm{b}}} E(0) = \sqrt{\frac{2\epsilon\epsilon_0}{e_0 n_{\mathrm{b}}} |\phi_s|} \tag{11.34}$$

where terms of the order of kT/e_0 have been neglected. For practical purposes it is convenient to express ϕ_s through the flat-band potential:

$$L_{\mathrm{sc}} = \sqrt{\frac{2\epsilon\epsilon_0}{e_0 n_{\mathrm{b}}} |(\phi - \phi_{\mathrm{fb}})|} \tag{11.35}$$

Problems

1. Consider the case of small band bending, in which $|e_0\phi(x)| \ll kT$ everywhere. Expand the exponential in Eq. (11.1), keeping terms up to first order, and calculate the distribution of the potential.
2. (a) Prove that $n_c p_v = N_c N_v \exp(-E_g/kT)$. (b) The effective densities of states N_c and N_v are typically of the order of 10^{19} cm^{-3}. Estimate the carrier concentrations in an intrinsic semiconductor with a band gap of $E_g = 1$ eV, assuming that the Fermi level lies at midgap.
3. Consider the interface between a semiconductor and an aqueous electrolyte containing a redox system. Let the flat-band potential of the electrode be $E_{\text{fb}} = 0.2$ V and the equilibrium potential of the redox system $\phi_0 = 0.5$ V, both versus SHE. Sketch the band bending when the interface is at equilibrium. Estimate the Fermi level of the semiconductor on the vacuum scale, ignoring the effect of dipole potentials at the interface.

References

1. A. Many, Y. Goldstein, and N.B. Grover, *Semiconductor Surfaces.* North Holland, Amsterdam, 1965.
2. V.A. Myalin and Yu.V. Pleskov, *Electrochemistry of Semiconductors.* Plenum Press, New York, NY, 1967.
3. H. Gerischer, *Z. Phys. Chem.* NF **26** (1960) 223; **27** (1961) 40, 48.

12

Selected experimental results for electron-transfer reactions

Innumerable experiments have been performed on both inner- and outer-sphere electron-transfer reactions. We do not review them here, but present a few results that are directly relevant to the theoretical issues raised in the preceding chapters.

12.1 Validity of the Butler–Volmer equation

The Butler–Volmer equation (9.13) predicts that for $|\eta| > kT/e_0$ a plot of the logarithm of the current versus the applied potential (Tafel plot) should result in a straight line, whose slope is determined by the transfer coefficient α. Because of the dual role of the transfer coefficient (see Sect. 9.2), it is important to verify that the transfer coefficient obtained from a Tafel plot is independent of temperature. We shall see later that proton- and ion-transfer reactions often give straight lines in Tafel plots, too, but the apparent transfer coefficient obtained from these plots can depend on the temperature, indicating that these reactions do not obey the Butler–Volmer law in the strict sense.

In order to test the temperature independence of the transfer coefficient, Curtiss et al. [1] investigated the kinetics of the Fe^{2+}/Fe^{3+} reaction on gold in a pressurized aqueous solution of perchloric acid over a temperature range from 25° to 75°C. In the absence of trace impurities of chloride ions, this reaction proceeds via an outer sphere mechanism with a low rate constant ($k_0 \approx 10^{-5}$ cm s^{-1} at room temperature). Figure 12.1 shows the slope of their Tafel plots, $d(\ln i)/d\eta$, as a function of the inverse temperature $1/T$. The Butler–Volmer equation predicts a straight line of slope $\alpha e_0/k$, which is indeed observed. Over the investigated temperature range both the transfer coefficient and the energy of activation are constant: $\alpha = 0.425 \pm 0.01$ and $E_{act} = 0.59 \pm 0.01$ eV at equilibrium, confirming the validity of the Butler–Volmer equation in the region of low overpotentials, from which the Tafel slopes were obtained.

W. Schmickler, E. Santos, *Interfacial Electrochemistry*, 2nd ed.,
DOI 10.1007/978-3-642-04937-8_12, © Springer-Verlag Berlin Heidelberg 2010

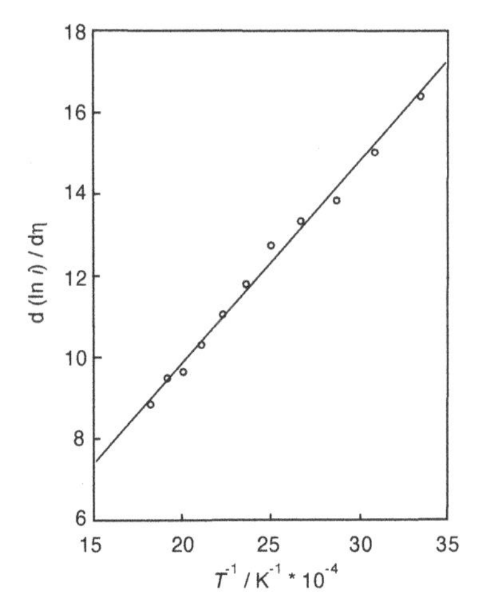

Fig. 12.1. Tafel slope as a function of the reciprocal temperature; reprinted with permission from [1].

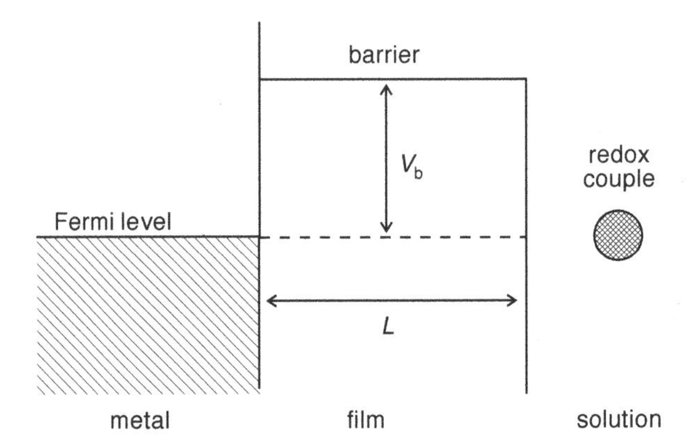

Fig. 12.2. Effective tunneling barrier for electron transfer in the presence of an insulating film (schematic).

12.2 Curvature of Tafel plots

The phenomenological derivation of the Butler–Volmer equation is based on a linear expansion of the Gibbs energy of activation with respect to the applied overpotential. At large overpotentials higher-order terms are expected to contribute, and a Tafel plot should no longer be linear. The theory presented in Chap. 10 makes a more detailed prediction: The current should become constant at high overpotentials. It is not easy to investigate this experimentally, because for large overpotentials the reaction is fast, and it is difficult to separate transport from kinetic effects.

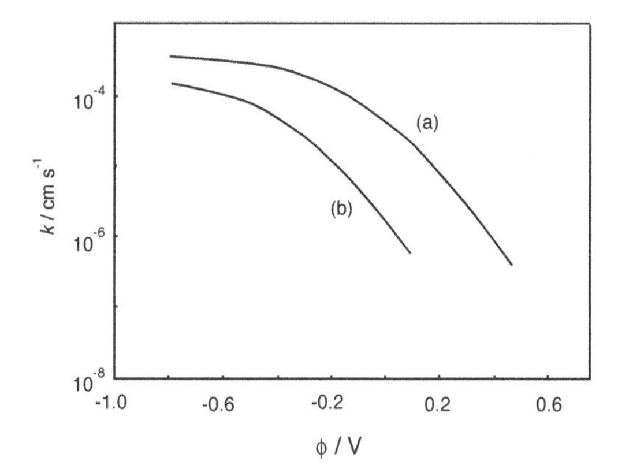

Fig. 12.3. Rate constants for the reduction of $[Mo(CN)_8]^{3-}$ (*upper curve*) and $[W(CN)_8]^{3-}$ (*lower curve*) on gold electrodes derivatized with a monolayer of $HO(CH_2)_{16}SH$. The electrode potential is given with respect to a Ag/AgCl electrode in saturated KCl. Data taken from [2].

The experiment is much easier to perform on electrodes coated with an insulating film, through which the transferring electron must tunnel, so that the reaction rate is decreased by several orders of magnitude. In a rough model a layer of intervening molecules can be represented by a rectangular barrier of a certain height V_b above the Fermi level, and a thickness L (see Fig. 12.2). According to the Gamov formula,[1] the probability $W(L)$ for an electron with an energy near the Fermi level to tunnel through the barrier is:

$$W(L) = \exp\left(-\frac{2}{\hbar}\sqrt{2mV_b}L\right) = e^{-\beta L} \tag{12.1}$$

where m is the electronic mass. Even though the effective barrier height is not well defined, a relation like Eq. (12.1) is often found to hold in practice, with decay constants β of the order of 1 Å^{-1}. The resulting reduction in the reaction rate makes it possible to measure the current at high overpotentials without running into the usual difficulty of transport limitations.

Miller and Grätzel [2] investigated a series of outer-sphere electron-transfer reactions on gold electrodes coated with ω-hydroxy thiol layers about 20 Å thick. They recorded current-potential curves over a range of 0.5–1 V, and found the expected curvature in all cases investigated. As examples we show the data for the reduction of $[Mo(CN)_6]^{3-}$ and $[W(CN)_8]^{-3}$ in Fig. 12.3; instead of the current these authors plotted the rate constants. The curves follow the theoretical equations (10.14) and (10.15) quite well. By a fitting procedure energy of reorganization can be obtained. In the case of $[Mo(CN)_6]^{3-}$ one obtains an energy of reorganization of about 0.4 eV. While this is a reasonable

[1] A derivation of this formula can be found in any textbook on quantum mechanics, e.g. Landau and Lifshitz [3].

value, some caution is required in the quantitative interpretation of such data: The effective barrier height V_b changes with the applied potential in a manner that is difficult to assess.

12.3 Adiabatic electron-transfer reactions

When a reaction is adiabatic, the electron is transferred every time the system crosses the reaction hypersurface. In this case the pre-exponential factor is determined solely by the dynamics of the inner- and outer-sphere reorganization. Consequently the reaction rate is independent of the strength of the electronic interaction between the reactant and the metal. In particular, the reaction rate should be independent of the nature of the metal, which acts simply as an electron donor and acceptor. Almost by definition adiabatic electron-transfer reactions are expected to be fast.

In order to investigate the dependence of a fast reaction on the nature of the metal, Iwasita et al. [4, 5] and Santos et al. [6] measured the kinetics of the $[Ru(NH_3)_6]^{2+/3+}$ couple on six different metals. Since this reaction is very fast, with rate constants of the order of 1 cm s^{-1}, a turbulent pipe flow method and the coulostatic method (see Chaps. 20 and 20, resp.) were used to achieve rapid mass transport. The results are summarized in Table 12.1; within the experimental accuracy both the rate constants and the transfer coefficients are independent of the nature of the metal. This remains true if the electrode surfaces are modified by metal atoms deposited at underpotential [5]. It should be noted that the metals investigated have quite different chemical characteristics: Pt and Pd are transition metals; Au, Ag, Cu are *sd* metals; Hg and the adsorbates Tl and Pb are *sp* metals. The rate constant on mercury involved a greater error than the others because the mercury film employed was stable only for a short time in the turbulent flow of the electrolyte. The anodic and cathodic transfer coefficients do not quite add up to unity; this was attributed to the slight curvature of the Tafel lines, an effect discussed previously.

12.4 Transition between adiabatic and non-adiabatic regime

As pointed out above, an intervening layer of adsorbates decreases the interaction between a redox couple and the electrode. Therefore, in the presence of a thick layer we expect electron transfer to be non-adiabatic, while on bare metals it is usually adiabatic. By systematically changing the thickness of the layer, one can pass from one limit to the other.

Metal	$k$cm^{-1} s^{-1}	α	β
Pt	1.2	0.39	0.47
Pd	1.0	0.46	0.44
Au	1.0	0.42	0.57
Cu	1.2	-	0.51
Ag	1.2	0.36	0.55
Hg	0.7 ± 0.2	0.44	0.52
Pt/Tl_{ad}	1.3	0.44	0.49
Pt/Pb_{ad}	1.1	0.36	0.48
Au/Tl_{ad}	1.0	0.49	0.42

Table 12.1. Rate constants and transfer coefficients of the $[Ru(NH_3)_6]^{2+/3+}$ couple on various metals [3, 4].

In the adiabatic case, the pre-exponential factor depends on the friction in the sense of Kramer's theory (see section 10.8 and Fig. 10.8). In the non-adiabatic case it should be determined by the electronic interaction and hence be independent of friction. These expectations were verified in experiments by Khoshtariya et al. [7], who studied electron transfer between a gold electrode and the $Fe(CN)_6^{3-/4-}$ couple in the presence of n-alkanethiol films with various numbers of n of methylene groups. At the same time, the viscosity of the solution was varied by adding glucose to the solution. In general a higher viscosity entails a higher friction, although there is no direct quantitative relation between the two quantities.

The results are shown in Fig. 12.4. For a thick intervening layer the rate is indeed independent of the viscosity, and falls off exponentially with the number of methylene units. In the absence of a layer, the rate decreases with the viscosity; for a thin layer, the rate depends weakly on the viscosity, indicating an interaction that lies between the limits of adiabatic and non-adiabatic transitions.

12.5 Electrochemical properties of SnO_2

Tin oxide is a semiconductor with a wide band gap of $E_g \approx 3.7$ eV, which can easily be doped with oxygen vacancies and chlorine acting as donor states. It is stable in aqueous solutions and hence a suitable material for n-type semiconducting electrodes.

The interfacial capacity follows the Mott-Schottky equation (11.2) over a wide range of potentials. Figure 12.5 shows a few examples for electrodes with various amounts of doping [8]. The dielectric constant of SnO_2 is $\epsilon \approx 10$; so the donor concentration can be determined from the slopes of these plots.

By extrapolating the Mott–Schottky plots to the abscissa, the flat band potential can be determined (see also Fig. 11.3). Its value depends on the donor concentration, as can be seen from the following argument. Consider

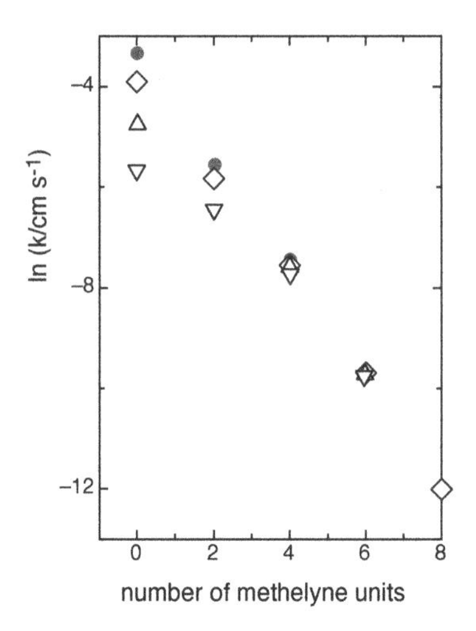

Fig. 12.4. Standard rate constant as a function of the number of methylene units for various solvent viscosities. The latter increases in the series: $\bullet < \diamond < \triangle < \triangledown$. Data taken from [7].

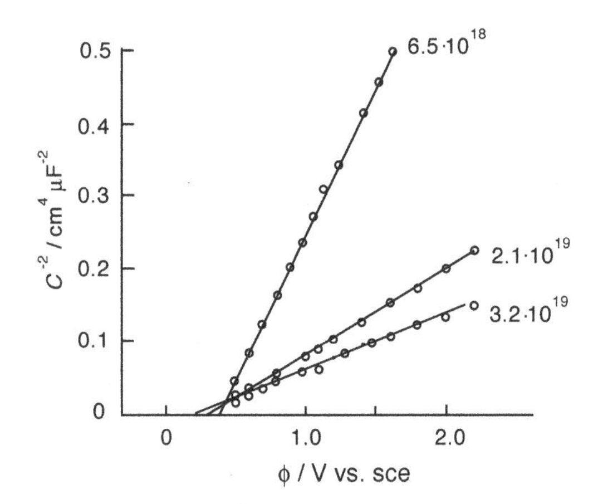

Fig. 12.5. Mott–Schottky plot for n-type SnO_2 for various donor concentrations. Data taken from [8].

two n-type semiconducting electrodes with different amounts of doping under depletion conditions (see Fig. 12.6). At a given electrode potential both electrodes have the same Fermi energy. The position of the band edges at the interface is fixed by the potential of the electrolyte solution, and is hence also the same for both electrodes. The position E_c of the conduction band depends on the donor concentration. The electrode with the higher concentration has

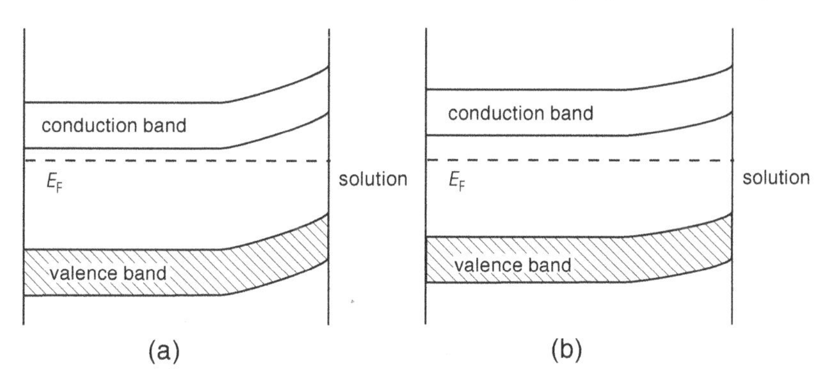

Fig. 12.6. Band bending at a SnO_2 semiconductor for two different donor concentrations. The semiconductor in (**a**) has the higher donor concentration; hence the Fermi level is closer to the conduction band, and the band bending is higher.

its conduction band closer in energy to the Fermi level, and thus shows a stronger band bending, and a lower value of the flat band potential.

As is often the case for metal-oxide electrodes in contact with aqueous solutions the surface of SnO_2 is covered by hydroxyl groups, which can dissociate according to the reactions:

$$\mathrm{SnOH} \rightleftharpoons \mathrm{SnO^-} + \mathrm{H^+} \tag{12.2}$$

$$\mathrm{SnOH} \rightleftharpoons \mathrm{Sn^+} + \mathrm{OH^-} \tag{12.3}$$

The equilibrium of these reactions depends on the pH of the solution. Changing the pH by one unit involves a change of 60 meV in the electrochemical potential. Since the amount of Sn at the surface is fixed, the equilibrium is shifted in such a way that the inner potential changes by 60 mV, which entails a corresponding shift of the band bending and hence of the flat-band potential.

Memming and Möllers [8] have investigated a series of redox reactions on doped SnO_2 electrodes. As is to be expected for an n-type semiconductor, most reactions proceed via the conduction band – the oxygen-evolution reaction, which occurs at high potentials and under strong depletion conditions, being an exception. Figure 12.7 shows current-potential curves for the Fe^{2+}/Fe^{3+} reaction for two different amounts of doping. For a low donor concentration the current follows the theoretical equation (11.14) for a conduction-band mechanism quite well. In particular the anodic current is almost constant for $\eta > kT/e_0$, while the cathodic branch shows a transfer coefficient of $\beta \approx 1$. However, on highly doped electrodes the current-potential curves are similar to those observed on metals. In this case the space-charge regions at the surface are so thin that the electrons can tunnel through them. Typically, the total width of the barrier is still too large for tunneling, but the electron can tunnel through the top of the barrier, as indicated in Fig. 12.8. So the position of the conduction band in the bulk is important. Figure 12.8 shows the conditions for the anodic reaction: Raising the overpotential increases the overlap between the density of reduced states and the conduction band and hence the anodic current density.

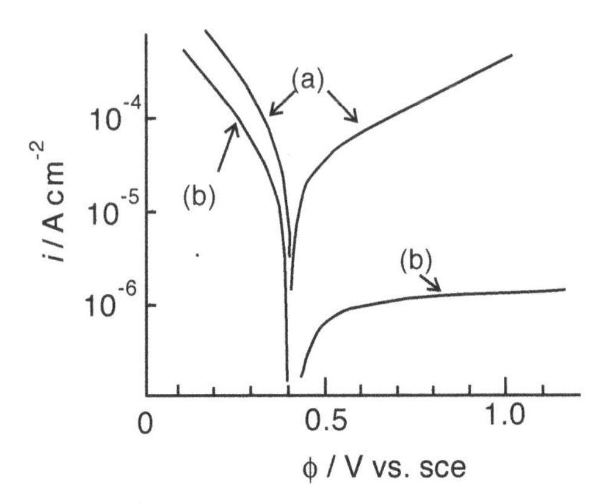

Fig. 12.7. Current-potential curves for 0.05 M Fe^{2+}/Fe^{3+} in 0.5 M H_2SO_4 at SnO_2 electrodes with two different donor concentrations; (**a**) 5×10^{19} cm^{-3}, (**b**) 5×10^{17} cm^{-3}. Data taken from [8].

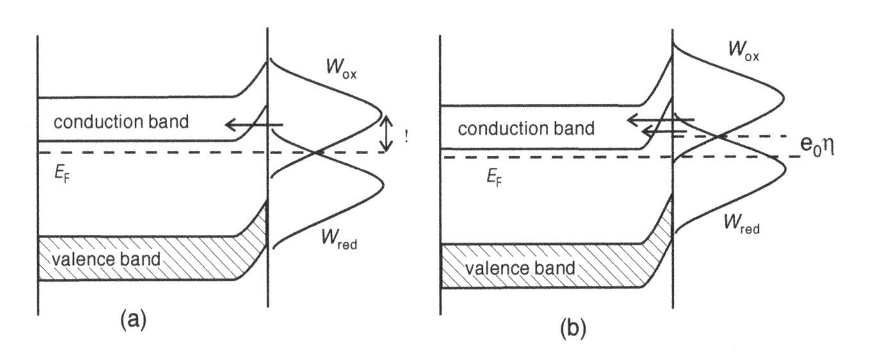

Fig. 12.8. Tunneling through the space-charge layer at equilibrium and for an anodic overpotential. Note that the band bending is stronger after the application of the overpotential. The *arrows* indicate electrons tunneling near the top of the space-charge barrier.

12.6 Photocurrents on WO_3 electrodes

In Sect. 11.4.1 we gave an outline of the photoeffects caused by electron-hole generation by photons with an energy above that of the band gap. An example is shown in Fig. 12.9, where the photocurrent generated in n-type semiconducting WO_3 is plotted for three different wavelengths of the incident light [9]. Note that Eq. (11.25) is obeyed better for larger wavelengths, for which the absorption coefficient α is smaller, and the relation $\alpha L_{sc} \ll 1$ is better fulfilled. In all cases considered Eq. (11.25) holds over a range of intermediate potentials, and the corresponding straight lines extrapolate to the flat-band potential.

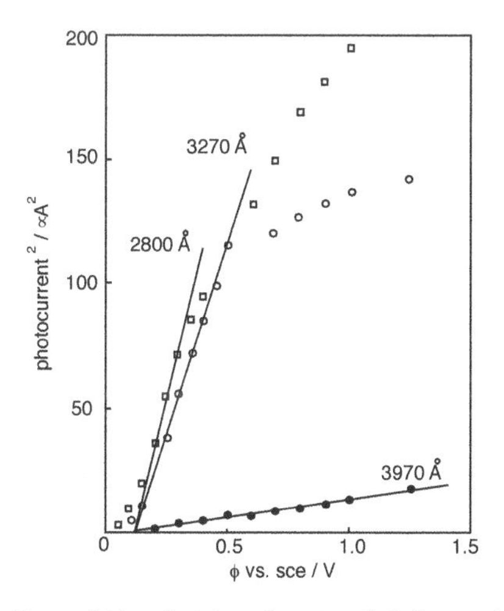

Fig. 12.9. Determination of the flat-band potential from the photocurrent. Data taken from [9].

When the light penetrates far into the semiconductor, minority carriers that are generated in the bulk can diffuse into the space-charge layer and contribute to the photocurrent. In this case Eq. (11.23) must be replaced by *Gärtner's equation* [10]:

$$j_p = e_0 I_0 \left(1 - \frac{\exp\left(-\alpha L_0 (\phi - \phi_{\text{fb}})^{1/2}\right)}{1 + \alpha L_p} \right) \tag{12.4}$$

where L_p is the *diffusion length* of the holes, which is the average distance that a hole travels before it disappears by recombination or by being trapped in a localized electronic state. A derivation is outlined in Problem 3. For $\alpha L_p \ll 1$ the contribution from the bulk is negligible, and Gärtner's equation reduces to Eq. (11.23).

Problems

1. From the Gamov formula, calculate the probabilities for an electron and for a proton to tunnel through barriers of 1 and 10 Å thickness with a height of 1 eV.
2. The anodic current density of a certain electron-transfer reaction on a film-covered electrode is found to be given by:

$$j_a = C \exp\left(-\frac{2}{\hbar}\sqrt{2mV_b}L\right) \exp\frac{\alpha e_0 \eta}{kT} \tag{12.5}$$

where C is a constant. The barrier height depends on the overpotential through:

$$V_b = V_0(1 + \zeta\eta) \tag{12.6}$$

where ζ is a constant. Assuming that $\zeta\eta \ll 1$, derive an expression for the apparent transfer coefficent and show that it depends on temperature.

3. Gärtner's equation can be derived by calculating that part of the photocurrent which comes from the bulk. The concentration $p(x)$ of holes obeys the following equation, which combines the familiar diffusion equation with a source and a loss term:

$$\frac{\partial p(x)}{dt} = g(x) + D\,\frac{\partial^2 p(x)}{\partial x^2} - \frac{p(x)}{\tau} \tag{12.7}$$

where the source term $g(x)$ is given by Eq. (12.5), and D is the diffusion coefficient of the holes. The last term accounts for the loss of holes due to recombination or trapping, and τ is the lifetime of the holes. We consider stationary conditions, so that $\partial p/\partial t = 0$. The concentrations of holes far from the surface is negligible; so $\lim_{x\to\infty} p(x) = 0$. If we make the simplifying assumption that all carriers which reach the space-charge region are immediately carried to the surface, the second boundary condition is $p(L_{\rm sc}) = 0$. Solve the differential equation using the ansatz:

$$p(x) = Ae^{-x/L_D} + Be^{-\beta x} \tag{12.8}$$

where A, B, and β are constants, and $L_D = (D\tau)^{1/2}$ is the diffusion length of the holes. The bulk contribution to the photocurrent is given by the diffusion current at $x = L_{\rm sc}$:

$$j_p^b = e_0 D\,\frac{dp(x)}{dx} \tag{12.9}$$

When this is added to the contribution for the space-charge region given by Eq. (11.23) one obtains Gärtner's equation.

References

1. L.A. Curtiss, J.W. Halley, J. Hautman, N.C. Hung, Z. Nagy, Y.J. Ree, and R.M. Yonco, *J. Electrochem. Soc.* **138** (1991) 2033.
2. C. Miller and M. Grätzel, *J. Phys. Chem.* **95** (1991) 5225.
3. L.D. Landau and E.M. Lifshitz, *Quantum Mechanics, Non-relativistic Theory*, Pergamon Press, Oxford, 1965.

4. T. Iwasita, W. Schmickler, and J.W. Schultze, *Ber. Bunsenges. Phys. Chem.* **89** (1985) 138
5. T. Iwasita, W. Schmickler, and J.W. Schultze, *J. Electroanal. Chem.* **194** (1985) 355.
6. E. Santos, T. Iwasita, and W. Vielstich, *Electrochim. Acta* **31** (1986) 431.
7. D. Khoshtariya, D. Dolidze, L.D. Zusman, and D.H. Waldeck, *J. Phys. Chem. A* **105** (2001) 1818.
8. R. Memming and F. Möllers, *Ber. Bunsenges. Phys. Chem.* **76** (1972) 475.
9. M.A. Butler, *J. Appl. Phys.* **48** (1977) 1914.
10. W.W. Gärtner, *Phys. Rev.* **116** (1959) 84.

13

Inner sphere and ion-transfer reactions

In outer sphere reactions, electron transfer occurs from a distance. Therefore the interaction between reactant and electrode is comparatively weak, which greatly simplifies the theoretical treatment. In this chapter we will consider reactions in which the reactants come in close contact with the electrode, so that chemical interactions become important.

13.1 Dependence on the electrode potential

We consider the transfer of an ion from the solution to the surface of a metal electrode; we leave out protons for the moment, and defer their treatment to a later chapter on electrocatalysis. Ion transfer is accompanied by a simultaneous discharge of the transferring particle by a fast, usually adiabatic, electron transfer. The particle on the surface may be an adsorbate as in the reaction:

$$\mathrm{Cl^-(sol)} \rightleftharpoons \mathrm{Cl_{ad}} + \mathrm{e^-(metal)} \tag{13.1}$$

In this case the discharge can be partial; that is, the adsorbate can carry a partial charge, as discussed in Chap. 6. Alternatively the particle can be incorporated into the electrode as in the deposition of a metal ion on an electrode of the same composition, or in the formation of an alloy. An example of the latter is the formation of an amalgam such as:

$$\mathrm{Zn^{2+}} + 2\mathrm{e^-} \rightleftharpoons \mathrm{Zn(Hg)} \tag{13.2}$$

The reverse process is the transfer of a particle from the electrode surface to the solution; usually the particle on the surface is uncharged or partially charged, and is ionized during the transfer.

Ions are much heavier than electrons. While electrons can easily tunnel through layers of solution 5–10 Å thick, and protons can tunnel over short distances, up to a few tenths of an Ångstrom, ions do not tunnel at all at room temperature. The transfer of an ion from the solution to a metal surface can

W. Schmickler, E. Santos, *Interfacial Electrochemistry*, 2nd ed.,
DOI 10.1007/978-3-642-04937-8_13, © Springer-Verlag Berlin Heidelberg 2010

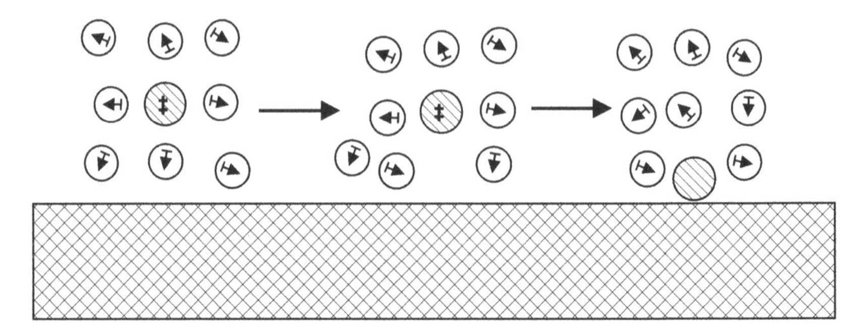

Fig. 13.1. Transfer of an ion from the solution onto the electrode surface (schematic).

be viewed as the breaking up of the solvation cage and subsequent deposition, the reverse process as the jumping of an ion from the surface into a preformed favorable solvent configuration (see Fig. 13.1).

In simple cases the transfer of an ion obeys a slightly modified form of the Butler–Volmer equation. Consider the transfer of an ion from the solution to the electrode. As the ion approaches the electrode surface, it loses a part of its solvation sphere, and it displaces solvent molecules from the surface; consequently its Gibbs energy increases at first (see Fig. 13.2). When it gets very close to the electrode, chemical interactions and image forces become large, and the Gibbs energy decreases again and reaches its minimum at the adsorption site. In addition, the ion experiences the electrostatic potential of the double layer. The total Gibbs energy curve has a maximum at a distance from the surface corresponding to about one diameter of the solvent molecules.

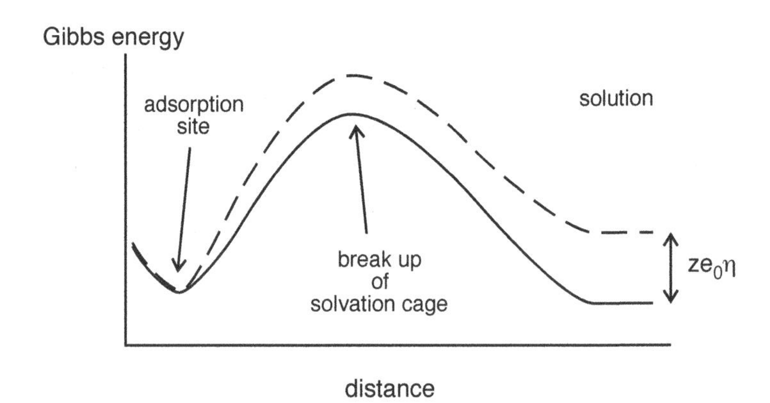

Fig. 13.2. Gibbs energy for the transfer of an ion from the solution to the electrode surface.

Application of an overpotential η changes the Gibbs energy for the ion transfer by an amount $ze_0\eta$, where z is the charge number of the ion. In addition, the double-layer field changes, and the structure of the solution may also be modified. This results in a change of the energy of activation by an amount $\alpha ze_0\eta$, where α is the transfer coefficient familiar from electron-transfer reactions.

These arguments are similar to those employed in the derivation of the Butler–Volmer equation for electron-transfer reactions in Chap. 9. However, here the reaction coordinate corresponds to the motion of the ion, while for electron transfer it describes the reorganization of the solvent. For ion transfer the Gibbs energy curves are less symmetric, and the transfer coefficient need not be close to 1/2; it may also vary somewhat with temperature since the structure of the solution changes.

The resulting potential dependence for the transfer of an ion to an adsorbed state is given by:

$$v = k_0 c_{\text{ion}}^s \exp \frac{\alpha zF(\phi - \phi_{00})}{RT} - k_0' \theta \exp \left(- \frac{(1-\alpha)zF(\phi - \phi_{00})}{RT} \right) \tag{13.3}$$

where c_{ion}^s is the concentration of the ion at the reaction site in the solution, and θ the coverage of the adsorbate. Since each ion carries a charge ze_0, the concomitant current density is $j = zFv$. If the concentration of the ions is unity, $c_{\text{ion}}^s = c^\ddagger$, and the electrode is at the standard equilibrium potential ϕ_{00}, the overall rate is zero by definition. Hence the exchange current density is:

$$j_{00} = zFk_0 c^\ddagger = zFk_0'\theta_{00} \tag{13.4}$$

where θ_{00} is the coverage at the standard equilibrium potential. The same equation can be used if the particle is incorporated into the surface of an electrode composed of the same material; in this case $\theta = 1$, formally.

If an adsorbed particle blocks a site for ion transfer, only a fraction $(1 - \theta)$ of the surface is available for the transfer, and we must replace Eq. (13.3) by:

$$\begin{aligned} j = {} & zFk_0 c_{\text{ion}}^s (1-\theta) \exp \frac{\alpha zF(\phi - \phi_{00})}{RT} \\ & -zFk_0'\theta \exp \left(- \frac{(1-\alpha)zF(\phi - \phi_{00})}{RT} \right) \end{aligned} \tag{13.5}$$

Thus, for a simple ion transfer a Butler–Volmer behaviour can be explained. But even in more complicated cases an empirical law of the form:

$$|\eta| = a + b \log_{10} \left(|j|/j^\ddagger \right) \tag{13.6}$$

is often found to hold both for high anodic and cathodic overpotentials; a and b are constants, and $j^\ddagger$ is the unit current density, which is introduced to make the argument of the logarithm dimensionless. This relation is known as

Tafel's law, and the coefficient b as the *Tafel slope*. It can be recast into the form:

$$j_a = j_0 \exp \frac{\alpha F \eta}{RT} \quad , \quad j_c = j_0 \exp \left(-\frac{\beta F \eta}{RT} \right) \tag{13.7}$$

for the two directions, but the two *apparent transfer coefficients* α and β need not be independent of temperature or even add up to unity. The experimental results for ion-, and also for proton-transfer reactions depend critically on the state and hence the preparation of the electrode surface, and different authors sometimes get different results. Sometimes Tafel's law is purely phenomenological. Many mechanisms may give rise to a rate that depends exponentially on the change in the reaction Gibbs energy.

13.2 Rate-determining step

Many ion-transfer reactions involve two or more steps. Often one of these steps proceeds more slowly than the others, and if the reaction proceeds under stationary conditions, this step determines the overall rate. We will elaborate this concept of a *rate-determining step* further. For this purpose consider a reaction taking place according to the general scheme:

$$\begin{aligned} \nu_1 A_1 + X_1 &\rightleftharpoons \mu_2 B_2 + X_2 \\ \nu_2 A_2 + X_2 &\rightleftharpoons \mu_3 B_3 + X_3 \\ &\text{up to} \\ \nu_{n-1} A_{n-1} + X_{n-1} &\rightleftharpoons \mu_n B_n + X_n \end{aligned} \tag{13.8}$$

This is a series of reactions, and the substances X_i $(i = 2, \ldots, n-1)$ are intermediates that are generated in one step and consumed in the next. The individual steps can be electrochemical or chemical reactions, or even mass-transport steps like the diffusion of a species from the bulk of the solution to the interface. The overall reaction is:

$$X_1 + \sum_{i=1}^{n-1} \nu_i A_i \rightleftharpoons X_n + \sum_{i=2}^{n} \mu_i B_i \tag{13.9}$$

When the reaction is stationary, all steps proceed at the same rate v, which is also the rate of the overall reaction. We denote by v_i and v_{-i} the rates at which the forward and backward reactions proceed. Then:

$$v = v_i - v_{-i} \tag{13.10}$$

Let step number j be *rate determining*; that is, its forward and backward rates are much smaller than those of the other steps:

$$v_j, v_{-j} \ll v_i, v_{-i}, \quad \text{for } i \neq j \tag{13.11}$$

Since $v = v_j - v_{-j}$, the overall rate is also much slower than those of the other steps:

$$v \ll v_i,\ v_{-i}, \qquad \text{for } i \neq j \tag{13.12}$$

so that all steps but the rate-determining one are in quasi-equilibrium:

$$v_i \approx v_{-i}, \qquad \text{for } i \neq j \tag{13.13}$$

Let k_i, k_{-i} $(i = 1, \ldots, n-1)$ denote the rate constants of the individual steps. The overall rate is then:

$$v = k_j [X_j][A_j]^{\nu_j} - k_{-j}[X_{j+1}][B_{j+1}]^{\mu_{j+1}} \tag{13.14}$$

where the square brackets denote concentrations. Since the other reactions are in equilibrium, the concentrations $[X_j]$ and $[X_{j+1}]$ can be calculated from the equilibrium constants $K_i = k_i/k_{-i}$. So the overall rate depends only on the rate constants of the rate-determining step and on the equilibrium constants of the other steps; the rate constants k_i, k_{-i} $(i \neq j)$ do not affect the reaction rate. This remains true if the reaction scheme involves parallel steps, but the rate-determining step can have no parallel step that is faster.

If one or more reaction steps involve charge transfer through the interface, their rates depend strongly on the applied potential. As the latter is varied, different steps may become rate determining. We will encounter examples in the remainder of this chapter.

13.3 Oxygen reduction

The electrochemistry of oxygen is of great technological importance. Oxygen reduction is used for energy generation in fuel cells and batteries, and it also plays a major role in corrosion. Oxygen evolution occurs in water electrolysis and a few other industrial processes. Unfortunately, both oxygen evolution and reduction are slow processes, and require a sizable overvoltage, of the order of several 100 mV, to proceed. This makes any fuel cell using oxygen inefficient. Thus, oxygen/hydrogen fuel cells deliver an open circuit voltage of the order of 0.8 V instead of the theoretical thermodynamic value of 1.2 V, and this cell voltage has not improved much since the invention of the fuel cell by Grove in 1830. The discovery of good and cheap catalysts for oxygen reduction is of paramount importance, if fuel cells are to play an important part in energy conversion.

The complete reduction of O_2 involves four electrons; in acid solutions the overall reaction is:

$$O_2 + 4H^+ + 4e^- \rightleftharpoons 2H_2O \tag{13.15}$$

and in alkaline solutions:

$$O_2 + 2H_2O + 4e^- \rightleftharpoons 4OH^- \tag{13.16}$$

Besides the complete reduction there is a competing process that stops at hydrogen peroxide. In acid solutions. the overall reaction is:

$$O_2 + 2H^+ + 2e^- \rightleftharpoons H_2O_2 \tag{13.17}$$

The overall reaction can be represented by the following scheme:

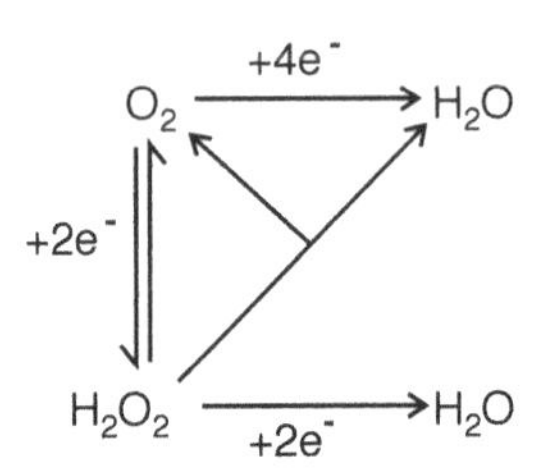

A direct pathway involving four electrons competes with an indirect pathway via H_2O_2. The intermediate H_2O_2 may escape into the solution or decompose catalytically into H_2O and O_2 on the electrode surface, so that the overall efficiency is greatly reduced.

The simultaneous transfer of four, or even of two, electrons is unlikely, and the overall reaction must contain several steps. A fair number of mechanisms have been proposed, but the experimental situation is quite unclear. Even for oxygen reduction on platinum in acid solutions, which is the best investigated case, there is disagreement about such basic facts as the value of the transfer coefficient or even the reaction order with respect to oxygen. The reaction is very sensitive to the structure of the electrode surface. For example, on Au(111) the reaction only delivers hydrogen peroxide, while on Au(100) the full reduction to water can be observed under favorable circumstances. Further, the reaction generally does not take place at a bare metal surface, but at an electrode that is at least partially covered with on oxygen species. Thus, at potentials above 0.75 V platinum is covered by an oxygen species. The first step in this process is probably:

$$H_2O \rightarrow OH_{ad} + H^+ + e^- \tag{13.18}$$

The adsorbed OH seems to undergo further transformation at higher potentials.

The presence of oxygen species has a great effect on oxygen reduction. Since the exact state of the surface depends on the preparation and on operating conditions, it is not surprising that the experimental situation is contradictory. Because of the absence of a clear picture, we limit ourselves to a few observations.

On most metals, the rate-determining step seems to be the transfer of the first electron, which can occur either by:

$$O_2 + e^- \rightarrow O_2^- \tag{13.19}$$

or

$$O_2 + e^- + H^+ \rightarrow O_2H_{ad} \tag{13.20}$$

In the first case, the O_2^- ion can only be a short-lived adsorbed intermediate, since it is not stable in the bulk solution. Both mechanisms result in a transfer coefficient of about 1/2 and a reaction order of unity with respect to oxygen, which are often, but not always, observed. Both initial steps are compatible with a variety of reaction sequences.

The development of DFT, and the rapid increase in computing power in recent years, has made it possible to investigate the *thermodynamics* of each step in a postulated sequence, provided it does not lead to a charged species like O_2^-, which is difficult to treat because of its strong interaction with the solvent. There is much activity in this area, and it is too early to pass judgement on any existing work and include it in a textbook. But in order to give an idea of what can be done, we take a brief look at a recent work by Nørskov et al. [1], which is presently much discussed. One of the mechanisms investigated by these authors is:

$$\frac{1}{2}O_2 \rightarrow O_{ad} \tag{13.21}$$

$$O_{ad} + H^+ + e^- \rightarrow HO_{ad} \tag{13.22}$$

$$HO_{ad} + H^+ + e^- \rightarrow H_2O \tag{13.23}$$

which is termed the dissociative mechanism because of the first step. Each step is a well-defined chemical reaction; the electrode potential enters into the energy of the electrons transferred. The authors calculated the reaction free energies – but not the activation energies – of these steps on a variety of metals and concluded, that generally the desorption of an adsorbed oxygen or hydroxyl limits the overall rate. This contradicts the findings that the first electron transfer according to Eq. (13.19) or (13.20) determines the rate. Another difficulty is that the energies of the intermediate states depend strongly on the state of the surface, especially on the presence of other adsorbates. Nevertheless, the authors explain the overall trends quite well. In any case, we believe that DFT-based calculations will play an important role in understanding oxygen reduction.

13.4 Chlorine evolution

In many ways the evolution of chlorine is the anodic analog of hydrogen evolution, which we will discuss in Chap. 14. The overall reaction is:

$$2Cl^- \rightleftharpoons Cl_2 + 2e^- \tag{13.24}$$

The standard equilibrium potential is 1.358 V vs. SHE and is thus a little higher than that for the oxygen reaction (1.28 V vs. SHE), so in aqueous solutions the two reactions generally proceed simultaneously. Chlorine production is a process of great industrial importance, and it is crucial to suppress oxygen evolution; in practice current efficiencies of 98% for chlorine evolution are achieved, because oxygen evolution is a slow process with a low exchange current density. In addition, the presence of chloride inhibits the formation of oxide films.

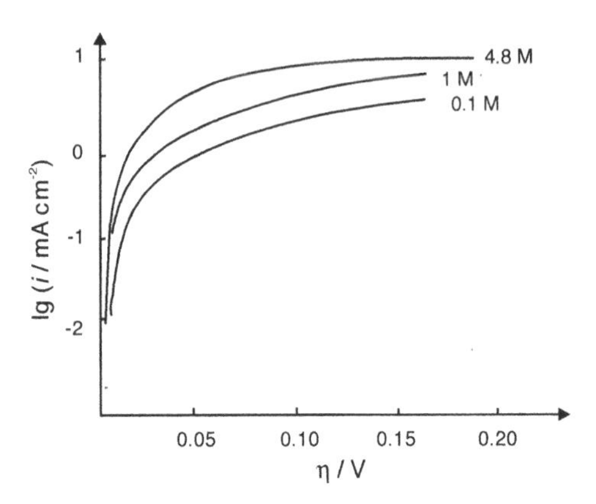

Fig. 13.3. Current-potential curves for chloride evolution on platinum from aqueous solutions. Data taken from [2].

The two main reaction mechanisms are analogous to the mechanisms for hydrogen evolution. The Volmer–Tafel mechanism is:

$$Cl^-(\mathrm{sol}) \rightleftharpoons Cl_{ad} + e^- \tag{13.25}$$

$$2\,Cl_{ad} \rightleftharpoons Cl_2(\mathrm{sol}) \tag{13.26}$$

while the Volmer–Heyrovsky mechanism corresponds to:

$$Cl^-(\mathrm{sol}) \rightleftharpoons Cl_{ad} + e^- \tag{13.27}$$

$$Cl_{ad} + Cl^- \rightleftharpoons Cl_2 + e^- \tag{13.28}$$

Which mechanism is observed in a particular situation depends on the electrode material. The reaction is well understood on platinum [2]. Usually platinum is covered with OH radicals at a potential of about 0.8 V vs. SHE, and at higher potentials an oxide film is formed. Though the formation of the oxide film is somewhat inhibited in the presence of Cl^-, a thin film is present in the potential region where chlorine is evolved. The presence of the film actually seems to catalyze the reaction, probably because it prevents the formation of a strong adsorption bond between Cl and Pt, which would slow down the desorption. At high overpotentials the current becomes constant (see Fig. 13.3); this indicates that the reaction proceeds according to the scheme of Eqs. (13.25) and (13.26) (Volmer–Tafel mechanism), and chemical desorption is the rate-determining step at high potentials.

Technical electrodes usually consist of a mixture of RuO_2 and TiO_2 plus a few additives. They are called *dimensionally stable anodes* because they do not corrode during the process, which was a problem with older materials. These two substances have the same rutile structure with similar lattice constants, but RuO_2 shows metallic conductivity, while pure TiO_2 is an insulator. The reaction mechanism on these electrodes has not yet been established; the experimental results are not compatible with either of the two mechanisms discussed above [2].

13.5 Oxidation of small organic molecules: methanol and carbon monoxide

The electrooxidation of small organic molecules is not as simple as one might assume. We choose as examples the oxidation of methanol and of carbon monoxide, because the former is a potential fuel for energy conversion, and the latter is involved in the poisoning of electrocatalyst.

Methanol is a small organic molecule easy to obtain, but its dehydrogenation involves several steps as can be appreciated from the mechanism proposed by Bagotzki [3].

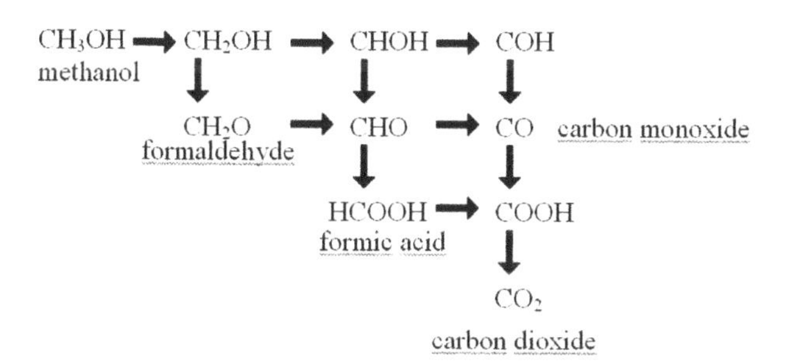

The first studies of the electrochemical oxidation of methanol were carried out by Müller et al. [4] in the nineteen twenties. Since that time, methanol has been considered as a promising candidate for fuel cells [5]. Because of its importance, the field is well reviewed in the literature (see for example, [3, 6, 7]). Methanol has a high specific energy capacity; its complete oxidation to CO_2 delivers six electrons, so that it should be possible to obtain about 0.85 Ah/g of energy:

$$CH_3OH + H_2O \rightarrow CO_2 + 6H^+ + 6e^- \qquad (13.29)$$

The thermodynamic potential is 0.02V, a value very close to that of the hydrogen oxidation reaction. In a fuel cell with the oxygen reduction as cathodic reaction, the overall process is:

$$CH_3OH + 3/2O_2 \rightarrow CO_2 + 2H_2O \qquad (13.30)$$

This yields a theoretical potential for the cell of 1.21 V.

However, methanol oxidation is relatively slow, even at highly active platinum electrodes. It is a complicated reaction with several steps. The formation of formic acid and formaldehyde have been detected. During the 1970s, Capon and Parsons [8] proposed a dual mechanism for the oxidation of small molecules with active and with poisoning intermediates. The direct pathway involves weakly adsorbed species, while during the indirect pathway a strongly adsorbed intermediate CO is formed, which inhibits further methanol oxidation. Thus, the catalysis of CO oxidation also becomes an important topic. In addition, it has been proposed [3] that under certain conditions, a possible weakly adsorbed intermediate COH_{ads} can age and transform to the inhibiting CO, too.

In order to investigate the electrooxidation of the strongly adsorbed poison, it is necessary to separate this process from those corresponding to the oxidation of the reactant diffusing from the bulk. In the seventies, Stonehart and Kohlmayr [9] employed a flux cell, which allows replacing, after the formation of the poisoning intermediate, the solution containing the active reactant by a fresh nitrogen saturated electrolyte. Then the electrooxidation of this species can be measured by a potentiostatic pulse or a potential sweep without any diffusional contribution. This procedure is more effective than removing the electrode from a solution and inserting into another cell. The potential is maintained under control during the whole experiment, and changes are also avoided in the adsorbate, since partial desorption and oxidation caused by contact with air are excluded. This simple technique was forgotten and again recovered in the nineteeneighties [10].

An interesting technique complementary to electrochemical measurements to investigate the nature of the intermediates is Differential Electrochemical Mass Spectrometry (DEMS) developed in the eighties at the University of Bonn [11, 12]. The mass signal of different products coming from the oxidation reaction can be followed on-line during the electrochemical process. The first study using isotope-marked material ($^{13}CH_3OH$ and ^{13}CO) was undertaken by Willsau and Heitbaum [13]. Other similar experiments in the group of Vielstich [14], using a flow cell to separate the contribution of the strongly adsorbed intermediate from the oxidation of methanol diffusing from the bulk confirmed that the adsorbate does not contain any methylic hydrogen. DEMS has been also employed to investigate the oxidation of CO. Here we show an example which illustrates the sensitivity of this method [15]. Figure 13.4 shows the current and the mass signal corresponding to the production of CO_2 measured simultaneously during the electrooxidation of CO adsorbed at 0.05 vs. RHE at Pt in 0.05 M $HClO_4$. The panels on the left show a potentiodynamic, those on the right a potentiostatic experiment. Although CO is a simple molecule, the results exhibit a multiplicity of processes. Particularly surprising is the correspondence between the decay at the beginning of the current transient with the detection of CO_2. One would have expected that at short times the current should mainly contain contributions due to double layer charging and

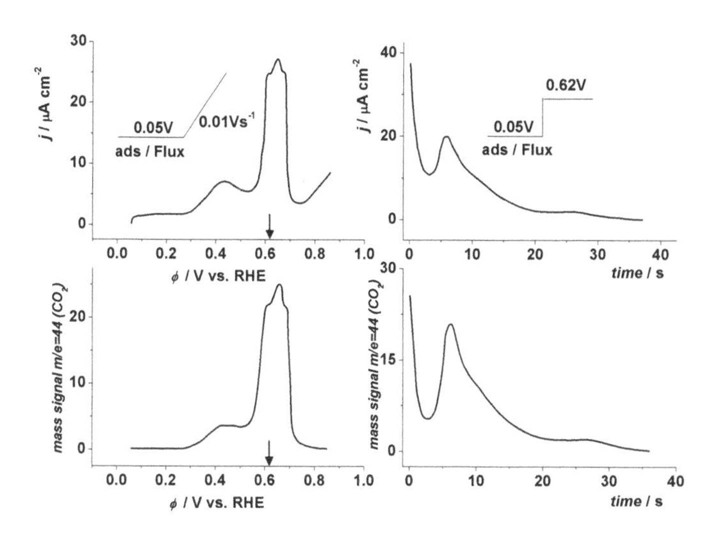

Fig. 13.4. Current (*upper panels*) and DEMS (*lower panel*) signals during CO oxidation. The panels on the *left* show potentiodynamic, those on the *right* potentiostatic experiments. Data taken from [15].

other secondary processes like the oxidation of traces of hydrogen formed at lower potentials. However, this process corresponds to the first broad peak observed at the potentiodynamic scan, as demonstrated by other experiment [15]. If the monolayer of CO is previously partially oxidized, the first peak at the scan and the decay at the mass transient disappear, while a current decay corresponding to the other processes mentioned above is still observed. The other conclusion that we can draw is that all processes involve the complete oxidation of the adsorbate to CO_2. Thus, we can disregard the assumption that the multiplicity is due to the formation of intermediate products such as formic acid or formaldehyde.

There is general agreement that the reaction steps involved in the oxidation of CO are those proposed by Gilman in the 1960s [16]. The oxidation reaction occurs between an adsorbed CO species and a surface-bonded OH species:

$$CO_{ads} + OH_{ads} \rightarrow CO_2 + H^+ + e^- \tag{13.31}$$

The multiplicity has been also attributed to the oxidation of CO adsorbed at different types of surface sites, and it strongly depends on the electrolyte composition (especially the type of anions and pH). The introduction of well-defined electrode surfaces by using single crystals during the eighties was the next milestone in order to understand the oxidation process of small molecules. We show here the first current transients obtained by a potential step to 0.62 V

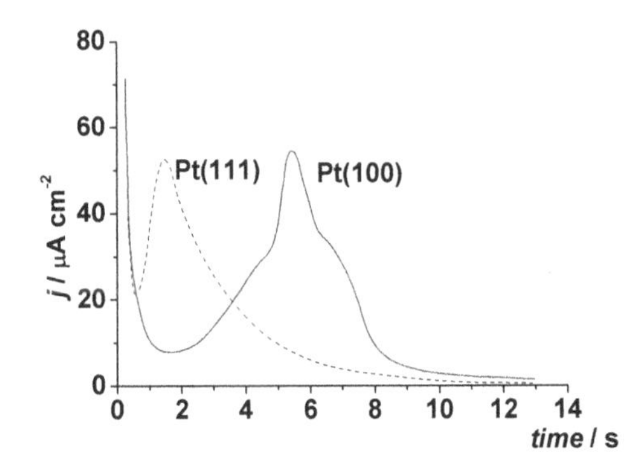

Fig. 13.5. Current transients for the oxidation of CO in 0.05 M $HClO_4$; first CO was adsorbed at 0.05 V vs. RHE, and then the potential was stepped to 0.62 V.

for the oxidation of CO previously adsorbed at platinum single crystal electrodes (see Fig. 13.5) at 0.05 V vs. RHE in 0.05 M $HClO_4$ [15]. Although at that time the quality of single crystals was still not perfect, the results show a clear difference between both surfaces. The asymmetry and the shoulder in the transients peaks can be attributed to the presence of defects. In a subsequent work [17] the authors performed a systematic analysis of the effect on the transient response by the introduction of perturbations in the surface. They observed an acceleration of the oxidation process; in addition the multiple oxidative behavior becomes more complex when the perturbations are larger. Later, Lai et al. [7] investigated the CO oxidation on stepped surfaces with (111) terraces of different sizes. They found that the rate of oxidation is proportional to the step density, and concluded that it takes place exclusively at the steps. They suggested that the mobility of CO on the (111) terraces must be high.

There is some disagreement about the mechanism for CO electrooxidation. Several authors describe the transients behavior on the basis of the Langmuir–Hinshelwood mechanism [7], while others [18] suggest a nucleation and growth mechanism of the oxide islands in the CO monolayer.

Returning to the electrooxidation of methanol, an important contribution to find a good catalyst was the introduction of bimetallic electrodes, particularly platinum–ruthenium system. Today, this synergy effect is the subject of many investigations (see, for example, [19]). An enhancement of the oxidation rate can occur if a modifier induces a decrease of the poisoning branch. This effect can be produced by different mechanisms: a third-body effect (surface sites are blocked for the poison), a bifunctional mechanism or a modification of the electronic properties. The bifunctional effect is believed to occur for the oxidation of adsorbed CO on Ru-modified Pt surfaces: adsorbed CO reacts with the oxygen containing species OH_{ads} adsorbed on neighboring sites, which is more abundant on Ru (or adsorbed at lower potentials on Ru) than on Pt. In the case of methanol (reacting to adsorbed CO) it is generally accepted that 3–4 Pt atoms are necessary for the accommodation of the methanol molecule. This is the reason for the inactivity of PtSn surfaces for methanol oxidation and also for the fact that Pt-Ru alloys with a low Ru content are best for methanol oxidation.

13.6 Comparison of ion- and electron-transfer reactions

At a first glance ion- and electron-transfer reactions seem to have little in common. In an ion-transfer reaction the reacting particle is transferred from the bulk of the solution through the solvent side of the double layer right onto the electrode surface, where it is adsorbed or incorporated into the electrode, or undergoes further reactions such as recombination. In contrast, in an outer-sphere electron-transfer reaction the reactant approaches the electrode up to a distance of a few Ångstroms, and exchanges an electron without penetrating into the double layer. In spite of these differences both types of reactions follow the same phenomenological Butler–Volmer law, at least for small overpotentials (i.e. up to a few 100 mV).

However, a closer inspection of the experimental data reveals several differences. For ion-transfer reactions the transfer coefficient α can take on any value between zero and one, and varies with temperature in many cases. For outer-sphere electron-transfer reactions the transfer coefficient is always close to 1/2, and is independent of temperature. The behavior of electron-transfer reactions could be explained by the theory presented in Chap. 10, but this theory – at least in the form we have presented it – does not apply to ion transfer. It can, in fact, be extended into a model that encompasses both types of reactions [20]; proton transfer reactions are special and will be treated in Chap. 14.

To construct such a unified model, we combine the theory of adiabatic electron transfer with the concept of desolvation, and calculate two-dimensional adiabatic free energy surfaces as a function of the solvent coordinate q and the distance from the surface. The details of such calculations are beyond the scope of this book, but the principles are easy to understand. We shall discuss three examples: an outer-sphere electron transfer, the adsorption of a simple ion, and the deposition of a divalent metal ion. The surfaces we present are by no means exact, but are sufficiently accurate to explain the qualitative differences and the trends.

With these preparations we can understand potential-energy surfaces that have been calculated for simple electron- and ion-transfer reactions. Figure 13.6 shows a free-energy surface for the Fe^{2+}/Fe^{3+} reaction as a function of both the distance x from the surface and the generalized solvent coordinate q. The calculations were performed for the equilibrium potential. At distances far from the electrode surface we observe two valleys, one for $q = -2$, which

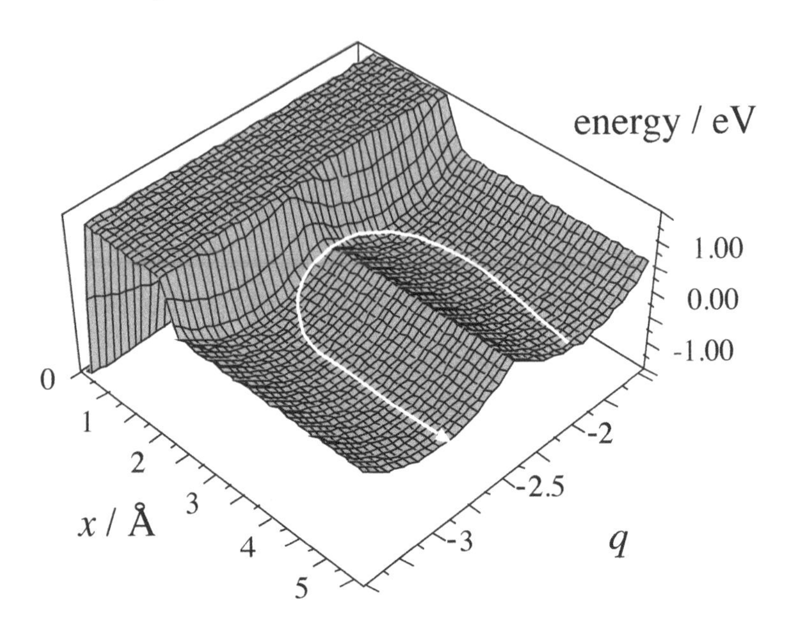

Fig. 13.6. Adiabatic free-energy surface for the Fe^{2+}/Fe^{3+} reaction. Close to the electrode surface the energy has been cut off at 1.5 eV for clarity; all energies are relative. The reaction path is indicated by the *dashed white line.*

corresponds to the Fe^{2+}, and one for $q = -3$ for the Fe^{3+}. These two valleys are separated by an energy barrier with a height of about 0.25 eV. The energy of reorganization of this couple is $\lambda \approx 1$ eV, so the barrier height is $\lambda/4$ in accord with the model presented in Chap. 10. If we take a cross-section at a constant distance x from the metal we obtain a free-energy curve similar to the one shown in Fig. 10.4 for the case of equilibrium. If we let the particle approach the electrode surface there is at first little change in the potential-energy surface until we reach the region in which the particle loses a part of its solvation sphere. Since the energies of solvation of the ions are very large (about 19.8 eV for Fe^{2+} and 50 eV for Fe^{3+}) this requires a large energy, and the potential-energy surface rises sharply by several electron volts in this region. In fact, this rise is so sharp that we had to cut off the energy so that the ridge between the two ions remains visible. Right at the surface the particle is adsorbed, and another local minimum occurs in this region.

In this situation it is highly unlikely that an Fe^{2+} or Fe^{3+} will be adsorbed on the electrode surface, since it would have to overcome a huge energy barrier. It is much easier for these particles to cross the much smaller energy barrier (about 0.25 eV) separating the reduced and the oxidized states by exchanging an electron with the metal. However, we have to bear in mind that the potential-energy surface that is shown corresponds to an adiabatic reaction. In reality the reaction will be adiabatic only at short distances x from the metal surface, where the electronic interaction with the metal is strong. At larger separations the reaction will be nonadiabatic: When the particle reaches the ridge it will cross over into the other valley only with a small probability, which decreases exponentially with the distance x.

Therefore the electron-transfer reaction from Fe^{2+} to Fe^{3+} proceeds along a reaction path like the one indicated in the figure. Note that the electron-transfer step itself occurs practically at a constant distance from the metal surface; the reaction coordinate is given by the solvent coordinate. This is the reason why the treatment presented in Chap. 10 is valid.

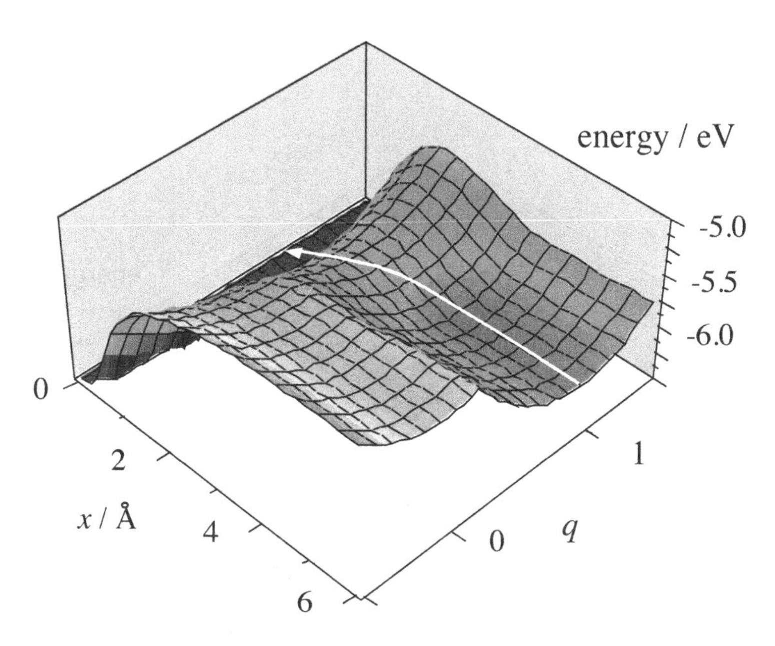

Fig. 13.7. Adiabatic free-energy surface for the adsorption of an iodide ion on Pt(100) at the pzc. The *white line* shows a possible reaction path.

As an example for an ion transfer reaction we consider the adsorption of an iodide ion on a Pt(100) surface. Figure 13.7 shows the potential-energy surface at the pzc. Far from the electrode we observe two valleys, one for the ion and one for the atom; both are separated by an energy barrier. As expected the energy of the ion is substantially lower than that of the atom (by about 0.65 V). Since the energy of the atom is so much higher it plays no role in the transfer of the ion, so we focus our attention on the latter. As the ion approaches the electrode surface it has to overcome an energy barrier in the region where it loses a part of its solvation sphere. Since the energy of solvation of the I^- ion is fairly small (about 2.5 eV) this energy barrier is comparatively low. Right on the electrode surface we observe another minimum, which corresponds to the adsorbed state. The reaction path for the ion transfer is indicated by the arrow in the figure. It is mainly directed towards the electrode surface, so the reaction coordinate is the distance of the ion from the electrode surface.

This potential-energy surface will change when the electrode potential is varied; consequently the energy of activation will change, too. These changes will depend on the structure of the double layer, so we cannot predict the value of the transfer coefficient α unless we have a detailed model for the distribution of the potential in the double layer. There is, however, no particular reason why α should be close to $1/2$. Also, a temperature dependence of the transfer coefficient is not surprising since the structure of the double layer changes with temperature.

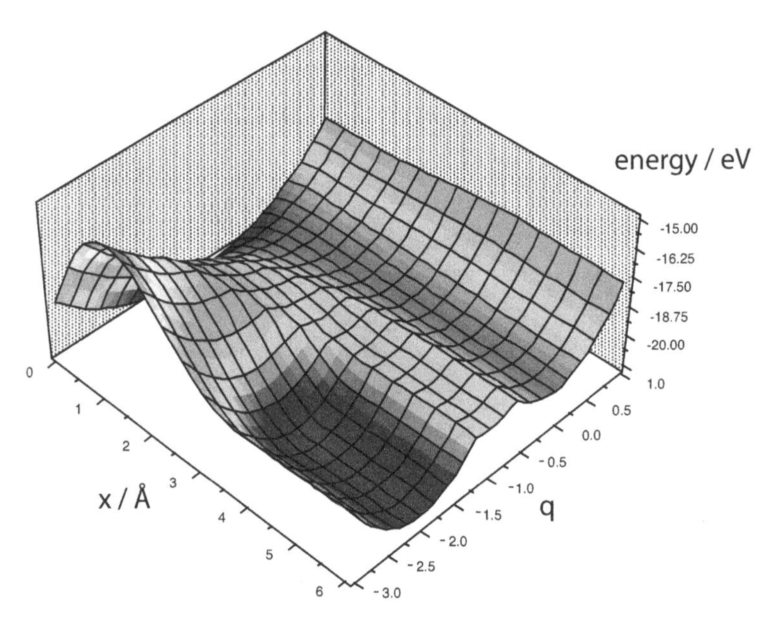

Fig. 13.8. Adiabatic free-energy surface for the deposition of a Zn^{2+} ion on mercury.

The behavior that we observed for the iodide ion is typical for the transfer of a univalent ion. For multivalent ions the situation is more complicated. Depending on the system under consideration and on the electrode potential a multivalent ion can either be transferred in one step, or its charge is first reduced by an electron-transfer reaction. As an example of the latter case we consider the deposition of a Zn^{2+} ion on mercury to form a zinc amalgam (Fig. 13.8). At large distances from the surface there are three valleys: a deep valley centered at $q = -2$ that corresponds to the Zn^{2+} ion, anther valley,

	Electron transfer	Ion transfer
Reaction coordinate	Solvent coordinate	Distance from surface
Transfer coefficient	$\alpha \approx 1/2$	$0 < \alpha < 1$
	Independent of T	May depend on T
Activation energy	Solvent reorganization	Solvent displacement

Table 13.1. Comparison of electron- and ion-transfer reactions

not so deep, representing the zinc atom, and in between there is a narrow, shallow valley for the Zn^+ ion. Remember that the passage from one valley to the other can occur only at short distances from the electrode. Because of the high energy of solvation of the doubly charged ion, the most favorable reaction path is via the valley for the Zn^+ ion, which is thus a short-lived intermediate.

Table 13.1 summarizes the different behavior of ion-transfer and electron-transfer reactions.

Problems

1. Consider a reaction consisting of an adsorption and an electron-transfer step:

$$A \rightleftharpoons A_{\rm ad} \tag{13.32}$$

$$A_{\rm ad} \rightleftharpoons A^+ + e^- \tag{13.33}$$

We ignore complications due to transport and assume that the surface concentrations of A and A^+ are constant. Let k_1 and k_{-1} denote the forward and backward rate constants of the adsorption reaction, so that the adsorption rate is given by:

$$v_{\rm ad} = k_1(1-\theta) - k_{-1}\theta \tag{13.34}$$

We assume that k_1 and k_{-1} are independent of the coverage and the electrode potential. We further assume that the rate of the electron-transfer step obeys a Butler–Volmer equation of the form:

$$v_{\rm et} = k_+\theta \exp\frac{\alpha F\eta}{RT} - k_-(1-\theta)\exp\left(-\frac{(1-\alpha)F\eta}{RT}\right) \tag{13.35}$$

where k_+ and k_- are constant. We have included the concentration of A^+ in k_- so that k_+ and k_- have the same dimensions. Assume that the reaction proceeds under stationary conditions. (a) Calculate the coverage at equilibrium and the exchange current density. (b) Derive the relation between current density and overpotential. (c) For small deviations from equilibrium derive a linear relation between current density and overpotential. (d) Derive simplified relations between current and potential for the cases where either the adsorption or the electron-transfer step are rate determining for all overpotentials, and sketch the corresponding Tafel plots.

2. In Chap. 4 we derived an expression for the work function of a simple redox reaction. Devise a suitable cycle to define the work function of the hydrogen evolution reaction. Check that it gives the correct order of magnitude for the absolute potential of this reaction.

References

1. J.K. Nørskov, J. Rossmeisl, A. Logadottir, L. Lindqvist, J.R. Kitchin, T. Bligaard, and H. Jonsson, *J. Phys. Chem. B* **108** (2004) 17886.
2. D.M. Novak, B.V. Tilak, and B.E. Conway, *Modern Aspects of Electrochemistry*, Vol. 14, edited by J. O'M. Bockris, B.E. Conway, and R.E. White. Plenum Press, New York, NY, 1982.
3. V.S. Bagotzky, Yu.B. Vassiliev, and O.A. Khazova, *J. Electroanal. Chem.* **81** (1977) 229.
4. E. Müller, *Z. Elektrochem. Ber. Bunsenges. Phys. Chem.* **28** (1928) 101.
5. A.N. Frumkin and B.I. Podlovchenko, *Dokl. Akad. Nauk SSR* **150** (1963) 349.
6. T. Iwasita and W. Vielstich, *Advances in Electrochemical Science and Engineering*, Vol. 1, edited by H. Gerischer and C.W. Tobas. VCH, New York, NY, 1990.
7. S.C.S. Lai, N.P. Lebedeva, T.H.M. Housmans, and M.T.M. Koper, *Top Catal.* **46** (2007) 320.
8. A. Capon and R. Parsons, *J. Electroanal. Chem.* **44** (1973) 1.
9. P. Stonehart and G. Kohlmayr, *Electrochim. Acta* **17** (1972) 369.
10. E. Santos and M.C. Giordano, *J. Electroanal. Chem.* **172** (1984) 201.
11. O. Wolter, M.C. Giordano, J. Heitbaum, W. Vielstich, W.E. O'Grady, P.N. Ross, and F.G. Will (eds.), Proceedings of the Symposium Electrocatalysis, Minneapolis, 1981, PV 82-2. The Electrochemical Society, Pennington, NJ, 1982, p. 235.
12. B. Bittins-Cattaneo, E. Cattaneo, P. Koenigshoven, and W. Vielstich, *Electroanalytical Chemistry*, Vol. 17, edited by A.J. Bard. M. Dekker, New York, NY, 1991.
13. J. Willsau and J. Heitbaum, *Electrochim. Acta* **31** (1985) 943.
14. T. Iwasita, W. Vielstich, and E. Santos, *J. Electroanal. Chem.* **229** (1987) 367.
15. E. Santos, E.P.M. Leiva, W. Vielstich, and U. Linke, *J. Electroanal. Chem.* **227** (1987) 199.
16. S. Gilman, *J. Phys. Chem.* **68** (1964)70.
17. E. Santos, E.P.M. Leiva, and W. Vielstich, *Electrochim Acta* **36** (1991) 555.
18. B. Love and J. Lipkowski, ACS Symposium Series 378, Soriaga Ed. Ch. 33 (1988).
19. H. Baltruschat, S. Ernst, and N. Bogolowski, *From Fundamental Aspects to Fuel Cell*, edited by E. Santos and W. Schmickler. Wiley, New York, NY, 2009.
20. W. Schmickler, *Chem. Phys. Lett.* **237** (1995) 152.

14

Hydrogen reaction and electrocatalysis

14.1 Hydrogen evolution – general remarks

The hydrogen evolution reaction is the most studied electrode process. Indeed, it has been suggested that focusing on hydrogen evolution has delayed the development of modern electrochemistry by years, if not decades [1]. However, in spite of all these efforts, the experimental data obtained by various groups do not agree all that well, and differences by one or two orders of magnitude for the rate constants are not unusual. In contrast to outer sphere reactions, where electron transfer occurs from a distance of a few Ångstroms, hydrogen evolution takes place right on the electrode surface, and is therefore highly sensitive to the state of the surface. Also, on transition metals like Pt or Ir the reaction is very fast, and hence difficult to investigate over a larger range of potentials.

The mechanism of hydrogen evolution and oxidation is simple. As we shall discuss below, there are only two different pathways, each consisting of two steps. Nevertheless, theoretical efforts at understanding the catalysis of this reaction were quite unsuccessful for a long time, and only in recent years we have begun to understand what makes a good catalyst. We shall consider it as a prototype to explain, how the rate of electron transfer depends on the electronic properties of the electrode.

It is instructive to consider the energetics involved in the reaction. If we start with the intact molecule H_2, we first have to break the bond. In the vacuum, the energy required for the breaking of the bond is about 4.5 eV. Taking the two electrons away to produce the protons requires twice the energy of ionization, 27.21 eV. The two electrons are transferred to the metal; in this transfer we gain twice the work function, i.e. an amount of the order of 9–11 eV. The remaining two protons are solvated. In water, the energy of hydration of the proton is about 11.5 eV – because its small size of the proton, it is the most strongly solvated ion. At the equilibrium potential, the potential drop between the metal and the solution balances the energy. From the large energies involved, it is obvious, that the hydrogen reaction cannot

W. Schmickler, E. Santos, *Interfacial Electrochemistry*, 2nd ed.,
DOI 10.1007/978-3-642-04937-8_14, © Springer-Verlag Berlin Heidelberg 2010

occur without catalysis: breaking a bond with an energy of 4.5 eV, or stripping away such a strongly attached solvation shell, requires substantial help from the metal.

14.2 Reaction mechanism

The overall reaction in acid media is:

$$2H^+ + 2e^- \rightleftharpoons H_2 \tag{14.1}$$

It is understood that the proton does not exist naked in the solution; the Zundel ion is the most likely precursor [2]. In alkaline media it proceeds according to:

$$2H_2O + 2e^- \rightleftharpoons H_2 + 2OH^- \tag{14.2}$$

In neutral solutions both reactions can occur.

We discuss acid solutions in greater detail. Two different mechanisms have been established. The first is the *Volmer–Tafel mechanism*, which consists of a proton-transfer step followed by a chemical recombination reaction:

$$H^+ + e^- \rightleftharpoons H_{ad} \quad \text{(Volmer reaction)} \tag{14.3}$$

$$2H_{ad} \rightleftharpoons H_2 \quad \text{(Tafel reaction)} \tag{14.4}$$

In the *Volmer–Heyrovsky mechanism* the second step also involves a charge transfer and is sometimes called *electrochemical desorption*:

$$H^+ + e^- \rightleftharpoons H_{ad} \quad \text{(Volmer reaction)} \tag{14.5}$$

$$H_{ad} + H^+ + e^- \rightleftharpoons H_2 \quad \text{(Heyrovsky reaction)} \tag{14.6}$$

On the *sp* metal like Hg and Cd the reaction seems to proceed via the Volmer–Tafel mechanism, with the Volmer reaction determining the rate. The reason is that the adsorption of the proton is endergonic on these metals, and the transfer of the proton with the entailing loss of solvation is not strongly catalyzed by the metal, effects which we shall discuss in greater detail below. On Ag and Cu, the second step is the Heyrowsky reaction. If the reaction is not too fast, the two mechanisms can be readily distinguished by their current transients [3]: When the (absolute value) of the current rises in time, the reaction proceeds via the Volmer–Heyrowsky mechanism, since at first a certain coverage with adsorbed hydrogen has be built up before the second electron can be transferred (see Fig. 14.1). In contrast, a decreasing current indicates a Tafel step, since the adsorbed hydrogen blocks sites for the Volmer step, and the recombination does not contribute to the current. On the *d* metals, both mechanisms can occur.

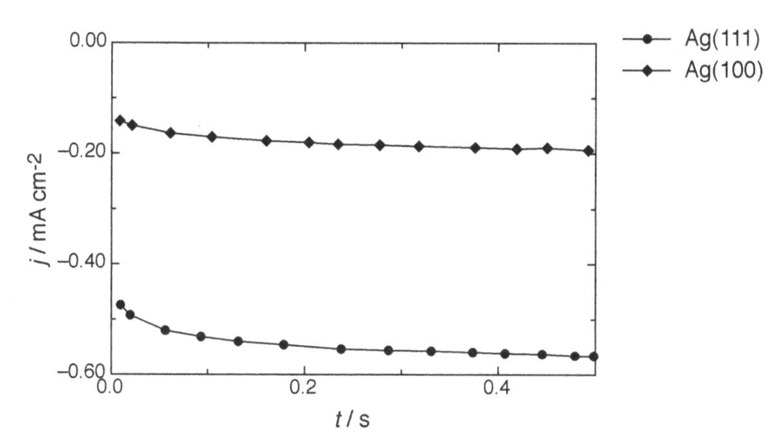

Fig. 14.1. Current transients for hydrogen evolution on Ag(100) and Ag(111) for an overpotential of −0.65 V vs. SHE in a solution of 0.1 M H_2SO_4 [4]; by convention, currents and overpotential for hydrogen evolution are negative. The rising absolute value of the current indicates the Heyrowsky mechanism.

14.3 Volcano plot

In the absence of a theory or model, it is natural to look for correlations in order to obtain at least a qualitative understanding. In the case of the hydrogen evolution reaction, more than ten different correlations [5] between the reaction rate and properties of the electrode were tried with limited success – amongst others, with the work function and with the presence of unfilled *d* orbitals. The best known of these correlations, and the only one that has survived, is the so-called volcano plot of the reaction rate, or the standard exchange current density, versus the energy of adsorption of a hydrogen atom on the electrode. Fig. 14.2 shows the version compiled by Trasatti [6]. Since at the time of the compilation reliable values for the adsorption energies were not available, the energy $E_{\mathrm{M-H}}$ for hydride formation was taken instead. Note that in this plot the formation energies are taken as positive, so that high values correspond to a high gain in energy.

Before discussion the data, let us look at the rationale behind this kind of plot, which is known as Sabatier's principle [7]. For high adsorption energies (low $E_{\mathrm{M-H}}$), the Volmer reaction is energetically uphill, which is unfavorable. With decreasing adsorption energy – increasing $E_{\mathrm{M-H}}$ – less energy is needed for the adsorption, and the reaction becomes faster. The optimum value should be close to a free energy of adsorption of zero. When the adsorption energy becomes more and more negative, the rate of the Volmer reaction still increases, but in the second step the energy of the initial state becomes lower and lower. So a stronger adsorption bond has to be broken either in the

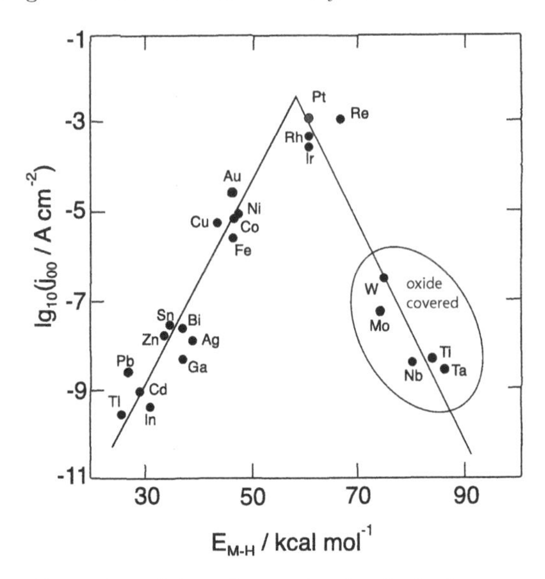

Fig. 14.2. Trasatti's version of the volcano plot for the hydrogen evolution reaction. Data taken from [6]

Heyrowski or the Tafel step, and thus the second step becomes slower and determines the rate [8, 9].

However, a look at the experimental evidence shows that things are not so simple. At first, with increasing $E_{\mathrm{M-H}}$, the rate does rise; there is also a descending branch, but under electrochemical conditions all metals on this branch are covered by an oxide film, whose presence impedes the reaction. This was not known at the time that this plot was proposed. If we leave the oxide-covered metals out, there is no evidence for a volcano plot, but only for the increasing branch.

With the advent of quantum chemical methods, in particular of DFT, it has become possible to calculate the free energy ΔG_{ad} of adsorption of the proton on the hydrogen scale, i.e. at the equilibrium potential, with an accuracy of about ± 0.1 eV. A modern version of the volcano plot, compiled by Nørskov et al. and with additions from ourselves, is shown in Fig. 14.3; the adsorption energies have been calculated in the absence of water, but similar calculations with water have shown that the effect of water on the adsorption is small. For gold and silver, there is a fair spread of experimental data; in these cases the extreme values have been indicated. On both metals, the higher values are more trustworthy than the lower ones, because they have been obtained on flame-annealed single crystals. Also, the rate varies somewhat between different single crystal planes of the same metal. Since all *sp* metals are bad catalysts, for this group the adsorption energy has only been calculated for Cd. Just as in Trasatti's plot, there is an overall tendency

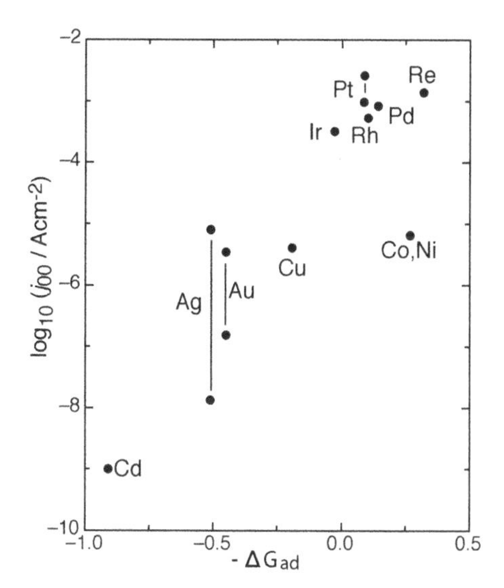

Fig. 14.3. Modern version of the volcano plot for the hydrogen evolution reaction. To ease the comparison with Fig. 14.2 the negative of the free energy of adsorption, $-\Delta G_{\rm ad}$, has been plotted on the x axis. Most data have been taken from [10]

for the rate to increase with decreasing $\Delta G_{\rm ad}$, but there is little evidence for a decreasing branch. Only cobalt and nickel have a low, negative energy of adsorption and small reaction rates. Note, that in Trasatti's plot these two elements lie on the ascending branch! Rhenium has an ever lower adsorption energy than Co and Ni, but the rate is about as high as on Pt.

There are only three metal on this plot with a highly exergonic adsorption energy: Ni, Co, Re. All other metals which adsorb hydrogen strongly are covered by oxide or hydroxide films under experimental conditions. So why is Re such a good catalyst, while Co and Ni are so bad? The answer is, that on transition metals with extended orbitals, like Pt, and Re, the strongly adsorbed hydrogen species does not participate in the reaction [11]. There is a second, weakly adsorbed species with a small positive value of $\Delta G_{\rm ad}$, which acts as the intermediate. We shall discuss this below in more detail. In contrast, Co and Ni have compact orbitals, and the weakly adsorbed species has an adsorption energy of about 1 eV and is thus highly unfavorable (Santos et al., Unpublished data). Therefore, the value of $\Delta G_{\rm ad}$ plotted on the x axis is relevant for Co and Ni, but not for Re, and not for Pt either.

We conclude that the idea underlying the volcano plot, Sabatier's principle, is sound, but there are complicating issues, such as the existence of several adsorption states, that one has to consider. Also, in the case of the coinage metals there is a compensating effect which makes them have roughly the same rates. We shall return to this point below.

14.4 Hydrogen evolution on Pt(111)

On the transition metals, hydrogen evolution is more complicated than on the *sp* and the coin metals. The rate determining step may change with potential or with the crystal face, and more than one species of adsorbed hydrogen may exist. As an example, we consider hydrogen evolution on Pt(111) from an aqueous solution in greater detail. Exactly the same mechanism has been observed on rhenium [12]. In these systems, one observes a cathodic transfer coefficient of about two, so that neither the Volmer nor the Heyrovsky step can be rate determining. We show, that it is consistent with the Volmer-Tafel mechanism, in which the Volmer reaction is fast, and the Tafel reaction is slow and rate determining.

Let us denote the rate constant for the Volmer reaction as $k_1(\eta)$, that of the back reaction as $k_{-1}(\eta)$. Since the Volmer reaction is fast and in quasi-equilibrium, we have:

$$k_1(\eta)c_p(1-\theta) = k_{-1}(\eta)\theta \tag{14.7}$$

where c_p denotes the surface concentration of H^+. At the equilibrium potential the coverage θ is determined by:

$$\frac{\theta}{(1-\theta)} = \frac{k_1(0)c_p}{k_{-1}(0)} = K_0 \tag{14.8}$$

At an arbitrary potential the equilibrium constant is

$$K = K_0 \ \exp\left(-F\eta/RT\right)$$

since the free energy of the reaction changes by $-F\eta$; hence:

$$\frac{\theta}{1-\theta} = K_0 \ \exp\left(-\frac{F\eta}{RT}\right) \quad \text{or} \quad \theta = \frac{K_0 \ \exp(-F\eta/RT)}{1 + K_0 \ \exp(-F\eta/RT)} \tag{14.9}$$

Denoting the forward rate constant for the Tafel reaction by k_2 and that for the back reaction by k_{-2}, we can write the current density in the form:

$$j = Fk_2\theta^2 - Fk_{-2}c_{H_2}(1-\theta)^2 \tag{14.10}$$

where c_{H_2} is the surface concentration of molecular hydrogen. The current vanishes at equilibrium, so that $k_{-2} = k_2K_0^2$. This gives the following expression for the current:

$$j = Fk_2K_0^2 \left(\frac{\exp(-2F\eta/RT)}{\left[1 + K_0 \ \exp(-F\eta/RT)\right]^2} - \frac{c_{H_2}}{\left[1 + K_0 \ \exp(-F\eta/RT)\right]^2} \right) \tag{14.11}$$

Experimental current-potential curves show Tafel behavior with an apparent cathodic transfer coefficient of two, provided the overpotential is sufficiently negative so that the back reaction can be neglected [13]. This suggests that the coverage θ of adsorbed hydrogen is small at all experimentally accessible potentials, so that $K_0 \exp(-F\eta/RT) \ll 1$.

Since the coverage with the adsorbed intermediate hydrogen is small, this species cannot be the strongly adsorbed hydrogen we discussed in the previous sections, which has an adsorption energy of $\Delta G_{ad} \approx -0.3$ eV. So there must be a second, weakly adsorbed species, which is sometimes referred to as *hydrogen deposited at overpotentials* H_{opd}. Most DFT calculations suggest, that the strongly adsorbed hydrogen, H_{upd}, is adsorbed in the threefold hollow sites, and the weakly adsorbed species on top, with $\Delta G_{ad} \approx 0.2$ eV. Thus, during hydrogen evolution, the surface of Pt(111) is always covered with a monolayer of strongly adsorbed hydrogen, which does not participate in the reaction. The presence of this hydrogen inhibits the recombination: While on bare Pt(111) the Tafel reaction is endergonic but proceeds practically without activation, the recombination of the weakly adsorbed species is exergonic, with an activation energy of the order of $0.4 - 0.7$ eV – the exact value depends on the coverage with strongly asorbed hydrogen. At a first glance, this high value seems puzzling, since it is of the same order of magnitude as that predicted for the Volmer reaction on Au(111) (see Fig. 14.8). However, for the recombination reaction the pre-exponential factor is of the order of a typical surface vibrational frequency, 10^{14} s^{-1}, while for the Volmer reaction it is about the same as for an outer sphere reaction, i.e. $10^{10} - 10^{11}$ s^{-1} .

14.5 Principles of electrocatalysis on metal electrodes

During the last few years we, the authors, with contributions from Koper [14], have developed a theory for the catalysis of electrochemical electron transfer reactions. We present our basic ideas below; at the time of writing there is no other theory.

In Chap. 10 we had presented the theory for outer-sphere electron transfer. An extension of this theory to electrocatalytic reactions requires several major modifications. The interaction of the reactant with the metal is much stronger in this case, and so one has to consider in greater detail the interaction with the electronic bands of the electrode. Also, just like in ion-transfer reactions, during the reaction the reactant approaches the electrode surface, which adds an extra dimension, the distance from the surface.

In outer sphere reactions, the interaction of the reactant with the metal is weak, of the order of $\Delta \approx 10^{-3}$ eV or less; therefore induces only a slight broadening of the valence orbital into a density of states, with negligible effect on the reaction rate. For catalytic reactions, the interaction is of the order of several eV, so we have to consider it in much greater detail.

All metals that are used as electrode materials have a wide *sp* band – as an example see Fig. 2.2. Since these bands are rather structureless near the Fermi level, where electron transfer happens, they give rise to a constant broadening Δ_{sp} in the same way as it happens for outer-sphere reactions, only the broadening is much larger because of the proximity to the surface, and is typically of the order of 0.5–1 eV. The *sp* bands behave much the same on all metals, and metals that have only *sp* bands near the Fermi level are extremely bad catalysts, as both versions of the volcano plot show. Catalysis is effected by the *d* bands, which are much narrower and have more structure.

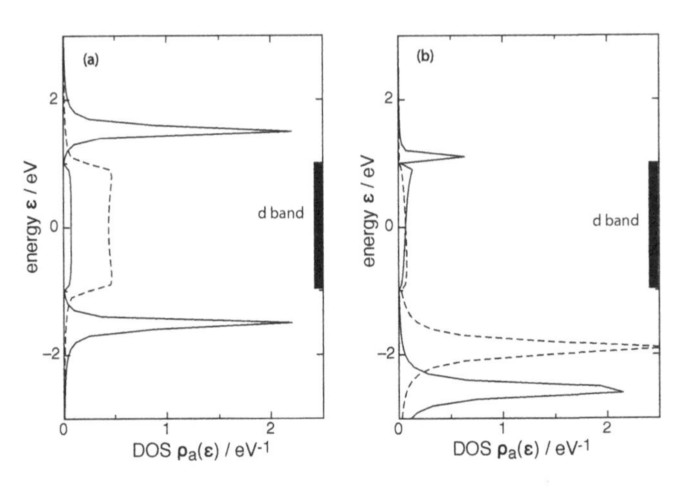

Fig. 14.4. Interaction of an orbital with a *d* band centered at $\epsilon = 0$; the position of the band is indicated at the *right* of both graphs. (**a**) Orbital centered at the *middle* of the band; *full line*: strong interaction; *dashed line*: weak interaction: (**b**) Orbital centered somewhat below the band; *full line*: strong interaction; *dashed line*: weak interaction

Let us consider the interaction of a reactant's orbital with a *d* band in some detail. Much can be learned from a simple model in which the *d* band is taken to have a semi-elliptic shape and a constant interaction, which does not depend on energy, with the reactant. Figure 14.4a shows the case in which the orbital is at the center of the band. If the interaction is weak, the orbital just gets broadened, and its DOS now extends over the width of the *d* band. However, if the interaction is strong something interesting happens: the DOS acquires two peaks, one on each side of the *d* band. The lower of these peaks has bonding character, the upper one is anti-bonding. This scenario is familar from the interaction between two atomic orbitals which combine to form bonding and anti-bonding molecular orbitals, only in this case one of the atomic orbitals has been replaced by a metal band. Note that the DOS also extends over the whole *d* band.

When the reactant's orbital lies somewhat below the *d* band, the separation into bonding and anti-bonding orbital also occurs for sufficiently strong interaction, but in this case the bonding peak is higher. This case is illustrated in Fig. 14.4b. If the orbital lies still lower, the effect of the d band becomes almost negligible. Mutatis mutandis, the same mechanism applies when the level lies above the d band.

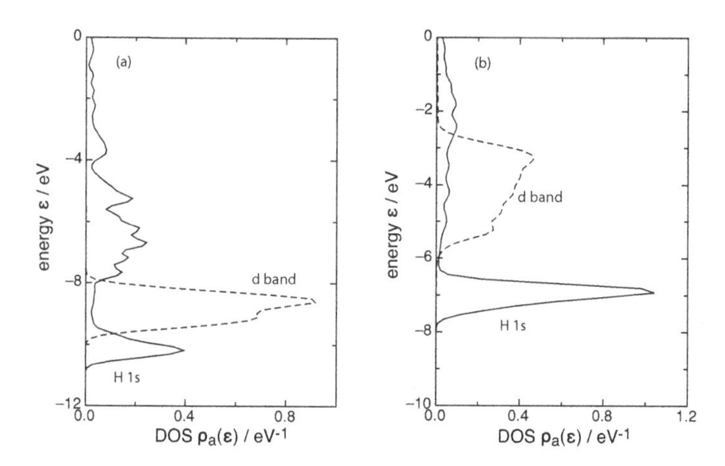

Fig. 14.5. 1s orbital of a hydrogen atom adsorbed on (**a**) Cd(0001) and (**b**) Ag(111); the d bands are shown as *dashed lines*; the Fermi level has been taken as the energy zero.

Corresponding results for the adsorption of a hydrogen atom obtained from DFT calculations are shown in Fig. 14.5. In the case of Cd(0001), the center of the hydrogen 1s orbital is near the center of the d band of Cd, and is split up into a bonding and an anti-bonding part. Since in Cd the d band lies well below the Fermi level, both the bonding and the anti-bonding part of the hydrogen orbital are filled, and so the d band does not contribute to the bonding. On the contrary, the Pauli repulsion with the d band weakens the bonding with the sp band; therefore the adsorption is weak and strongly endergonic at SHE (see Fig. 14.3). In the case of hydrogen on Ag(111), the 1s orbital lies below the d band, and the peak of the anti-bonding part of the hydrogen DOS is small. Again, both bonding and anti-bonding parts are filled, and no bonding results from the interaction with the d band.

With this preparation, we can explain how a d band situated near the Fermi level, and interacting strongly with the reactant, can greatly reduce the energy of activation and hence catalyse the reaction. The underlying mechanism, which was proposed by us [16], is illustrated in Fig. 14.6 for the case where a reaction of the type: $A \rightarrow A^+ + e^-$ is in equilibrium. In the initial state, with $q = 0$, the reactant's orbital is at $\epsilon = -\lambda$, situated well below the Fermi level, somewhat broadened by the interaction with the sp band, but in the situation depicted hardly influenced by the d band, which lies too far

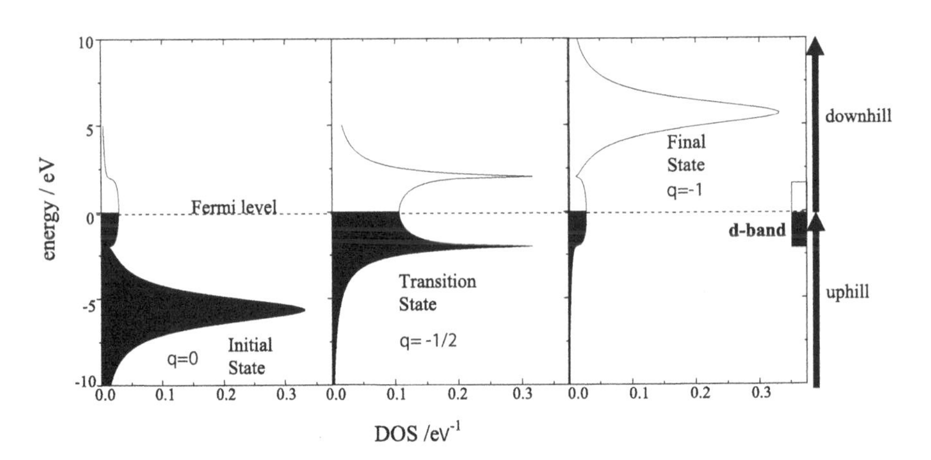

Fig. 14.6. Mechanism of electrocatalysis by a d band near the Fermi level.

above. During the course of the reaction, a thermal fluctuation of the solvent raises the electronic level of the reactant. The critical phase occurs, when this level passes the Fermi level of the metal. As indicated in the figure, a d band situated close to the Fermie level and interacting strongly with the reactant, induces a substantial broadening of the density of states. Since the electronic energy is given by the integral:

$$\int_{-\infty}^{0} \epsilon \rho_a(\epsilon) d\epsilon \tag{14.12}$$

the part that lies below the Fermi level substantially reduces the energy. In addition, the broadening entails that the occupation of the orbital, which is given by:

$$n = \int_{-\infty}^{0} \rho_a(\epsilon) d\epsilon \tag{14.13}$$

becomes less than unity as the system approaches the saddle point. Therefore the reactant becomes partially charged, and the interaction with the solvent further reduces the energy. Thus, a broadening of the DOS as the system passes the Fermi level lowers the energy of activation and thereby catalyses the reaction. This mechanism immediately explains why good metal catalysts, such as platinum or ruthenium, generally have a high density of d states near the Fermi level. However, the mere presence of these states is not enough, they must also interact strongly with the reactant to be effective.

Our theory, which combines electron transfer theory with DFT, makes it possible to perform calculations for hydrogen evolution on different metal surfaces. Figure 14.7 shows the density of states of the hydrogen $1s$ orbital at the saddle point of the Volmer reaction $H^+ + e^- \rightarrow H$ for two different catalysts,

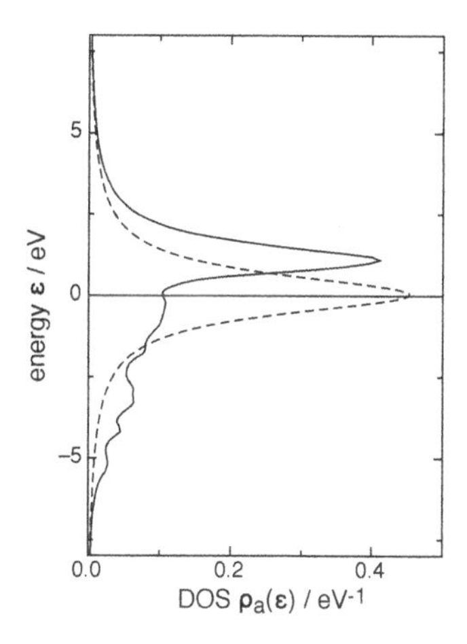

Fig. 14.7. Density of states of the hydrogen 1s orbital at the saddle point of the Volmer reaction. *Full line*: Pt(111), *dashed line*: Cd(0001).

Pt(111) and Cd(0001). On platinum, the DOS is substantially broadened at lower energies due to the strong interaction of hydrogen with the d band, while on cadmium the DOS is narrower and not affected by the d band, which lies too low. Also, on platinum the hydrogen carries a greater charge at the saddle point than on cadmium, so that the solvation is stronger. These two effects lead to a substantial lowering of the energy of activation.

14.6 Free energy surfaces for the Volmer reaction

From our theory we can calculate free energy surfaces for the hydrogen evolution reaction. Although at the time of writing the results are somewhat preliminary, they are certainly qualitatively correct, and at least they give the correct trends and orders of magnitude for the energy of activation.

On the *sp* metals and on the coin metals the Volmer reaction determines the rate at least at short times. On many transition metals, ΔG_{ad} is negative on the SHE scale (see Fig. 14.3), so that a monolayer of hydrogen is adsorbed at potential above SHE; in analogy with metal deposition, this effect is sometimes called *underpotential deposition of hydrogen*. As discussed above, the reaction then sometimes proceeds via a weakly adsorbed intermediate.

Since the Volmer reaction is of special interest, we show corresponding free energy surfaces for Pt(111) and Au(111) in Fig. 14.8 as representative examples. The free energy is plotted as a function of the solvent coordinate q and the distance d from the surface. On both surfaces, we see a minimum at $q = -1$ and large distances corresponding to the proton in solution – this is really the beginning of a valley centered at $q = -1$ extending into the bulk of

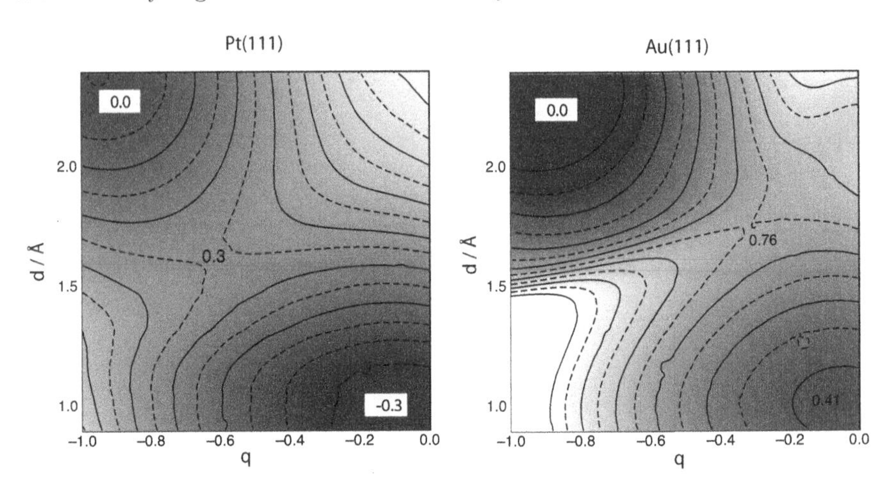

Fig. 14.8. Free energy surface for the Volmer reaction $H^+ + e^- \rightarrow A$ on Pt(111) (*left*) and on Au(111) (*right*); all energies are in eV.

the solution. By definition, the free energy corresponding to this state is zero on the SHE scale. A second minimum is centered at $q = 0$ and $d \approx 0.9$ Å; this represents the adsorbed hydrogen atom. On Pt(111), the free energy ΔG_{ad} of the adsorbed atom is negative, on gold it is positive (cf. Fig. 14.3). The two minima are separated by a barrier, whose saddle point gives the free energy of activation. On Pt, the barrier for the reaction is quite low, on Au it is sizable.

Similar surfaces have been calculated for a series of metals, and the following picture emerges: The *sp* metals behave like Cd; the *d* band plays no role, the *sp* bands do not differ much. The Volmer reaction is highly endergonic, and the energies of activation are of the order of 1 eV. On the coinage metals the *d* band does catalyse the reaction a little. The rate of the Volmer reaction is about the same on all three, due to a compensation between two competing effects: The free energy of adsorption increases from Cu to Au down the column of the periodic table because of Pauli repulsion, but the interaction constants increase in the same order due to the increasing size of the orbitals. On the early transition metals like Pt and Re, the reaction is fast, but various mechanisms may operate, sometimes via weakly adsorbed intermediates. In some cases the mechanism even depends on the crystal surface.

Problems

1. Consider the Volmer–Heyrovsky mechanism. Assume that the two reaction occur only in the forward direction, and that the coverage is so small that $(1 - \theta) \approx 1$. Derive the transient for the total current. In particular, show that the absolute value of the current rises, and discuss how the contributions from the two reactions can be separated.

2. Calculate the current transient for the Volmer–Tafel mechanism. As in Problem 1, consider only the forward direction, but this time the coverage need not be small. Show that the absolute value of the current decrease, and discuss how the rates for the two reactions can be determined separately.
3. Sketch the current potential curves according to Eq. (14.11) both for positive and negative overpotentials.
4. Consider a reactant orbital that passes the Fermi level, situated at $E_F = 0$, which is broadened into a Gaussian density of states of the form:

$$\rho(\epsilon) = \frac{1}{w\sqrt{\pi}} \exp(-x^2/w^2) \tag{14.14}$$

Calculate the electronic contribution:

$$\int_{-\infty}^{0} \epsilon\, \rho(\epsilon)\, d\epsilon \tag{14.15}$$

References

1. J.O'M. Bockris und A.K.N. Reddy, *Modern Electrochemistry*. Plenum Press, New York, NY, 1970.
2. F. Wilhelm, W. Schmickler, R.R. Nazmutdinov, and E. Spohr, *J. Phys. Chem. C* **112** (2008) 10814.
3. H. Gerischer and W. Mehl, *Ber. Bunsenges. Phys. Chem.* **59** (1955) 1049.
4. D. Eberhardt, Ph.D. Thesis, Ulm University, 2001.
5. O.E. Petri and G.A. Tsirlina, *Electrochim. Acta* **39** (1994) 1739.
6. S. Trasatti, *J. Electroanal. Chem.* **39** (1972) 163.
7. F. Sabatier, *La catalyse en chimie organique*. Berauge, Paris 1920
8. R. Parsons, *Trans. Faraday Soc.* **54** (1958) 1053.
9. H. Gerischer, *Bull. Soc. Chim. Belg.* **67** (1958) 506.
10. J.K. Norskøv, T. Bligaard, A. Logadottir, J.R. Kitchin, J.G. Chen, S. Pandelov, and U. Stimming, *J. Electrochem. Soc.* **152**, J23 (2005).
11. G. Jerkiewicz, *Prog. Surf. Sci.* **57** (1998)137.
12. M.J. Joncich and L.S. Stewart, *J. Elcctrochem.* **112** (1965) 717.
13. F. Ludwig, R.K. Sen, and E. Yeager, *Elektrokhimiya* **13** (1973) 847.
14. E. Santos, M.T.M. Koper, and W. Schmickler, *Chem. Phys. Lett.* **419** (2006) 421.
15. E. Santos and W. Schmickler, *ChemPhysChem* **7** (2006) 2282; E. Santos and W. Schmickler, *Chem. Phys.* **332** (2007) 39; E. Santos and W. Schmickler, *Angew. Chem. Int. Ed.* **46** (2007) 8262.

15

Metal deposition and dissolution

15.1 Morphological aspects

On a liquid metal electrode all surface sites are equivalent, and the deposition of a metal ion from the solution is conceptually simple: The ion loses a part of its solvation sheath, is transferred to the metal surface, and is discharged simultaneously; after a slight rearrangement of the surface atoms it is incorporated into the electrode. The details of the process are little understood, but it seems that the discharge step is generally rate determining, and the Butler–Volmer equation is obeyed if the concentration of the supporting electrolyte is sufficiently high. For example, the formation of lithium and sodium amalgams [1] in nonaqueous solvents according to:

$$\begin{aligned} Li^+ + e^- &\rightleftharpoons Li(Hg) \\ Na^+ + e^- &\rightleftharpoons Na(Hg) \end{aligned} \tag{15.1}$$

obeys the Butler–Volmer equation with transfer coefficients that depend on the solvent. On the other hand, the deposition of multivalent ions may involve several steps. As discussed in Chap. 13, the formation of zinc amalgam from aqueous solutions, with the overall reaction:

$$Zn^{2+} + 2e^- \rightleftharpoons Zn(Hg) \tag{15.2}$$

occurs in two steps: First, Zn^{2+} is reduced to an intermediate Zn^+ in an electron transfer step, and then the univalent ion is deposited [2].

In contrast, the surface of a solid metal offers various sites for metal deposition. Figure 15.1 shows a schematic diagram for a crystal surface with a quadratic lattice structure. A single atom sitting on a flat surface plane is denoted as an *adatom*; several such atoms can form an *adatom cluster.* A *vacancy* is formed by a single missing atom; several vacancies can be grouped to *vacancy clusters.* Steps are particularly important for crystal growth, with *kink* atoms, or atoms in the *half-crystal position*, playing a special role. When

W. Schmickler, E. Santos, *Interfacial Electrochemistry*, 2nd ed.,
DOI 10.1007/978-3-642-04937-8_15,

a metal is deposited onto such a surface, the vacancies are soon filled. However, the addition of an atom in the kink position creates a new kink site; so at least on an infinite plane the number of kink sites does not change, and the current is maintained by incorporation into these sites. Similarly the removal of a kink atom creates a new kink site. In the limit of an infinitely large crystal, the contribution of other sites can be neglected. For this reason Nernstian equilibrium is established between the ions in the solution and atoms in the half-crystal position.

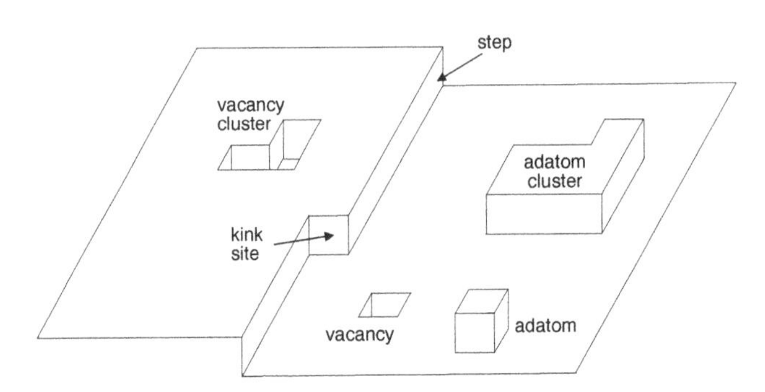

Fig. 15.1. A few characteristic features on a metal surface.

There are two different pathways for metal deposition: direct deposition from the solution onto a growth site, or the formation of an adatom with subsequent surface diffusion to an edge. Both mechanisms seem to occur in practice. If direct deposition is the dominant mechanism, the Butler–Volmer equation holds, provided the concentration of the supporting electrolyte is sufficiently high to eliminate double-layer effects. From our discussion above, it appears that metal deposition and growth can be viewed as a propagation of steps. On a perfect but finite metal plane any propagating step must at some time reach the edge, and the growth sites disappear. In this case a new nucleus for growth must be formed, a process that will be considered in the following. However, real crystals have *screw dislocations* (see Fig. 15.2), which propagate indefinitely, forming spiral structures.

15.2 Surface diffusion

If the dominant mechanism of deposition involves the formation of adatoms followed by surface diffusion to steps, the relation between current and electrode potential becomes complicated. The essential features can be understood within a simple model, in which we consider two parallel steps on the surface, a distance L apart (see Fig. 15.3. The surface diffusion of the adatoms is now a one-dimensional problem. Let $c_{\text{ad}}(x)$ be the surface concentration of the adatoms, and $D_{\rm ad}$ their diffusion coefficient. At equilibrium, $c_{\rm ad}(x) = c^0_{\rm ad}$ everywhere, the deposition and dissolution of adatoms balance and are characterized by an exchange current density $j_{0,\rm ad}$. A consideration of mass balance gives the following equation for the adatom concentration:

$$\frac{\partial c_{\rm ad}}{\partial t} = D_{\rm ad}\,\frac{\partial^2 c_{\rm ad}}{\partial x^2} + s(x) \tag{15.3}$$

where the source term $s(x)$ denotes the number of adatoms deposited at the position x per time and area. If the deposition and dissolution of the adatoms obey the Butler–Volmer equation, we have:

$$s(x) = \frac{j_{0,\rm ad}}{zF}\,\exp\left(-\frac{(1-\alpha)ze_0\eta}{kT}\right) - \frac{j_{0,ad}c_{\rm ad}}{zFc^0_{\rm ad}}\,\exp\,\frac{\alpha ze_0\eta}{kT} \tag{15.4}$$

The incorporation of the adatoms at the steps should be fast because no charge transfer is involved; hence the adatom concentration should attain its equilibrium value:

$$c_{\rm ad}(0) = c_{\rm ad}(L) = c^0_{\rm ad} \tag{15.5}$$

Under stationary conditions $\partial c_{\rm ad}/\partial t = 0$, and an ordinary differential equation results with Eq. (15.5) as boundary conditions, which can be solved explicitly by standard techniques. The resulting expression for the current density is:

$$j = j_{0,\rm ad}\left[\exp\,\frac{\alpha ze_0\eta}{kT} - \exp\left(-\frac{(1-\alpha)ze_0\eta}{kT}\right)\right]\frac{2\lambda_0}{L}\,\tanh\,\frac{L}{2\lambda_0} \tag{15.6}$$

where

$$\lambda_0 = \left(\frac{zFD_{\rm ad}c_{\rm ad}}{j_{0,\rm ad}}\right)^{1/2}\exp\left(-\frac{\alpha ze_0\eta}{2kT}\right) \tag{15.7}$$

λ_0 has the meaning of a penetration length of surface diffusion. We can distinguish two limiting cases:

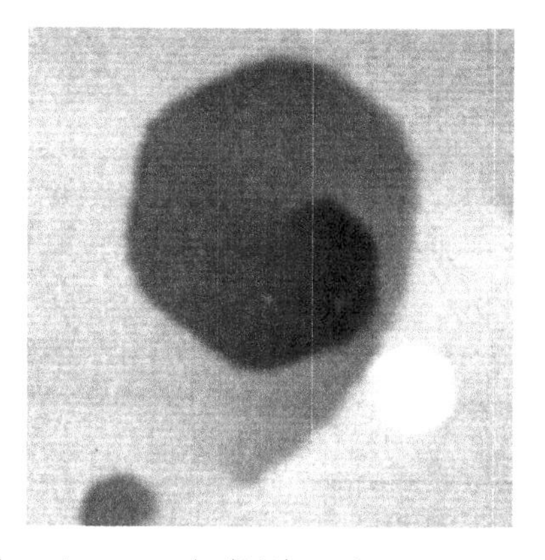

Fig. 15.2. Screw dislocation on a Ag(111) surface; courtesy of M. Giesen, Jülich.

1. $\lambda_0 \gg L$: the two terms involving L/λ_0 cancel, surface diffusion is fast, the deposition of adatoms is rate determining, and Eq. (15.6) reduces to the Butler–Volmer equation.
2. $\lambda_0 \ll L$: surface diffusion plays a major role, and the current density is:

$$j = j_{0,\mathrm{ad}} \left[\exp \frac{\alpha z e_0 \eta}{kT} - \exp \left(- \frac{(1-\alpha) z e_0 \eta}{kT} \right) \right] \frac{2\lambda_0}{L} \tag{15.8}$$

Substituting λ_0 from Eq. (15.7) gives:

$$j = \frac{2}{L} \left(\frac{zFD_{\mathrm{ad}} c_{\mathrm{ad}}^0}{j_{0,\mathrm{ad}}} \right)^{1/2} \left[\exp \frac{\alpha z e_0 \eta}{2kT} - \exp \left(- \frac{(1-\alpha) z e_0 \eta}{2kT} \right) \right] \tag{15.9}$$

which has the same form as the Butler–Volmer equation, but the apparent transfer coefficients are only half as large as those for the deposition and dissolution of the adatoms. Of course, real metal surfaces do not consist of steps running parallel and equidistantly from each other. However, even in the general case we would expect the kind of deviations from simple Butler–Volmer behavior as seen in Eq. (15.9), in particular a change in the apparent transfer coefficients.

15.3 Nucleation

A metal surface that is uniformly flat offers no sites for further growth. In this case a new nucleus, or center of growth, must be formed. Since small clusters of metal atoms consist mainly of surface atoms, they have a high energy content, and their formation requires an extra energy. The basic principles of

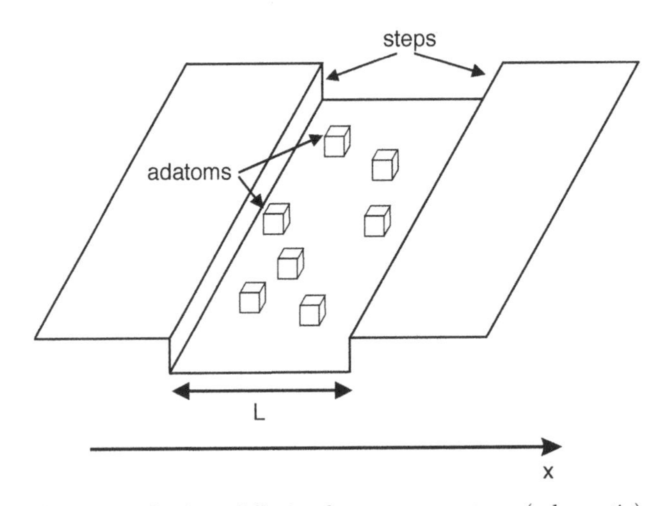

Fig. 15.3. Surface diffusion between two steps (schematic).

the formation of new nuclei can be understood within a simple model. We consider a small three-dimensional cluster of metal atoms on a flat surface of the same material, and suppose that the cluster keeps its geometrical shape while it is growing. A cluster of N atoms has a surface area of:

$$S = aN^{2/3} \tag{15.10}$$

where a is a constant depending on the shape of the cluster and the particle density n. For a hemispherical cluster of radius r the number of particles is:

$$N = \frac{2}{3}\pi r^3 n \tag{15.11}$$

so that:

$$a = (2\pi)^{1/3}\left(\frac{3}{n}\right)^{2/3} \tag{15.12}$$

The surface energy of a cluster is γS, where γ is the surface energy per unit area. For a liquid metal γ is identical to the surface tension. The electrochemical potential of a particle in the cluster contains a surface contribution, which is obtained by differentiating the surface energy with respect to N. Therefore:

$$\tilde{\mu} = \tilde{\mu}_\infty + \frac{2}{3}\gamma a N^{-1/3} \tag{15.13}$$

$\tilde{\mu}_\infty$ is the electrochemical potential for an atom in an infinite crystal. The Gibbs energy required to form a cluster by deposition from the solution is:

$$\Delta G(N) = N(\tilde{\mu}_\infty - \tilde{\mu}_s) + \gamma a N^{2/3} \tag{15.14}$$

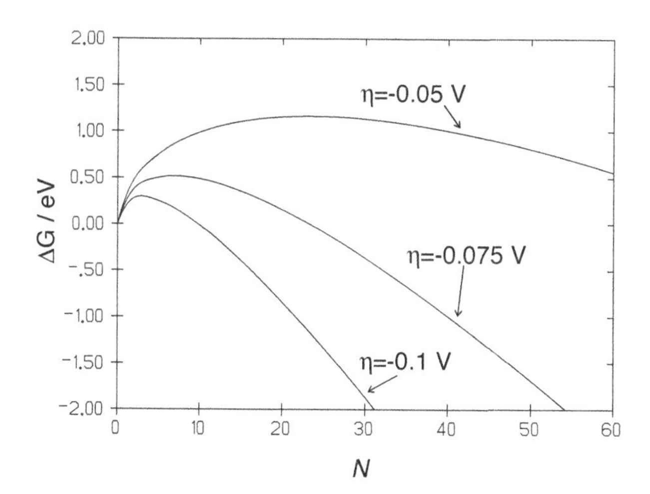

Fig. 15.4. Gibbs energy for the formation of a nucleus as a function of the particle number for various overpotentials.

where $\tilde{\mu}_s$ is the electrochemical potential of the metal ion in the solution. At equilibrium the two electrochemical potentials are equal, $\tilde{\mu}_\infty = \tilde{\mu}_s$; therefore, on application of an overpotential η, the difference $(\tilde{\mu}_\infty - \tilde{\mu}_s)$ is given by the product of the charge ze_0 of the metal ion and η; hence:

$$\Delta G(N) = Nze_0\eta + \gamma a N^{2/3} \tag{15.15}$$

Metal deposition can occur only if η is negative; so the Gibbs energy of a cluster as a function of the particle number N first rises, reaches a maximum, and then decreases. This is illustrated in Fig. 15.4 for three different overpotentials. Notice how strongly the curve depends on the applied overpotential. ΔG reaches its maximum for a critical particle number of:

$$N_c = -\left(\frac{2\gamma a}{3ze_0\eta}\right)^3 \tag{15.16}$$

where it takes on the value:

$$\Delta G_c = \frac{4(\gamma a)^3}{27(ze_0\eta)^2} \tag{15.17}$$

Clusters with a smaller number of particles than N_c will tend to dissolve, while larger clusters will tend to grow further. However, cluster formation and growth are stochastic processes, and there is a certain probability that subcritical clusters will grow, and supercritical clusters can still disappear. The Gibbs energy ΔG_c of a critical cluster is also the Gibbs energy of activation required to form a new nucleus for further crystal growth. The larger the absolute value of the applied (negative) overpotential, the higher the rate of nucleation. Once a nucleus has formed, it will continue to grow even if the overpotential is lowered.

While our arguments are simplified in several respects – three-dimensional clusters will not all have the same shape, and the use of a macroscopic concept like the specific surface energy γ is not really warranted – they are qualitatively correct, and Eqs. (15.16) and (15.17) are useful estimates.

15.4 Initial stages of deposition

We consider in greater detail the morphological aspects of metal deposition on a foreign substrate. Starting from a flat metal substrate S, on which metal atoms A are deposited; there are several principle mechanisms [3] – see Fig. 15.5:

1. The interaction of the atoms A with each other is stronger than with the substrate S. In this case, three-dimensional clusters are formed from the beginning. Obviously, this cannot take place at underpotentials. On the contrary, nucleation usually requires an overpotential for deposition to occur. This mechanism is known as Volmer–Weber or three-dimensional island growth (case (a) in Fig. 15.5).

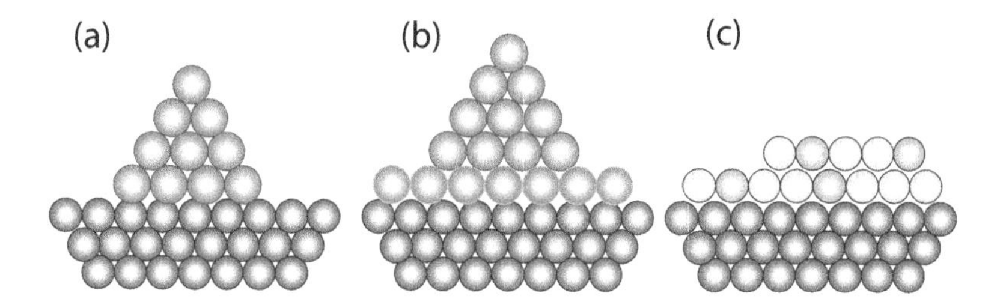

Fig. 15.5. Three growth modes: (**a**) Three-dimensional island growth (Volmer–Weber); (**b**) Stranski–Krastanov growth: (**c**) layer-by-layer or Frank-van-der-Merwe growth.

2. If the interaction of the deposited atoms A with the substrate S is stronger than with its own kind, a monolayer of A os S can be deposited at underpotential. There are two subcases:
   - If there is a considerable mismatch in the lattice structures of A and S, the first layer has a different, often incommensurate structure. Subsequently, three-dimensional clusters are formed. This is denoted as Stranski–Krastanov growth mode (case (b) in Fig. 15.5).
   - If there is no large mismatch between the crystallographic structures of the two metals A and S, a commensurate monolayer is formed. Subsequent layers are also epitaxic and deposited layer-by-layer; this is also known as the Frank-van-der-Merwe growth mode. After two or three layers have been deposited, the influence of the substrate is negligible, and the deposition proceeds in the same way as on the bulk metal A (case (c) in Fig. 15.5).
3. The deposited atoms A make a rapid place exchange with the substrate, and a surface alloy is formed. A well-known example is the deposition of nickel on Au(111) – see Fig. 15.6.

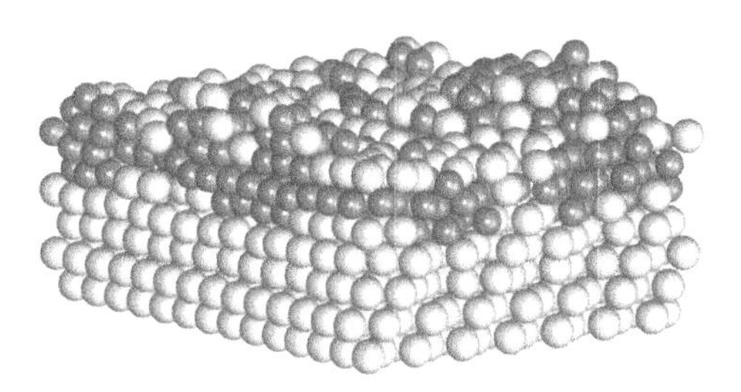

Fig. 15.6. Surface alloy formation during the deposition of nickel (*dark grey*) on Au(111). Result of a computer simulation [4].

Usually, the substrate is not flat, and the deposition may begin at steps or island. As an example, we consider the deposition of silver on Au(111), a system that exhibits underpotential deposition, in the vicinity of an island. At potentials well above the potential for bulk deposition, the rim of the gold island is decorated by silver atoms. With increasing potential the silver atoms spreads to the terrace, and finally deposition takes place on the whole surface – see Fig. 15.7.

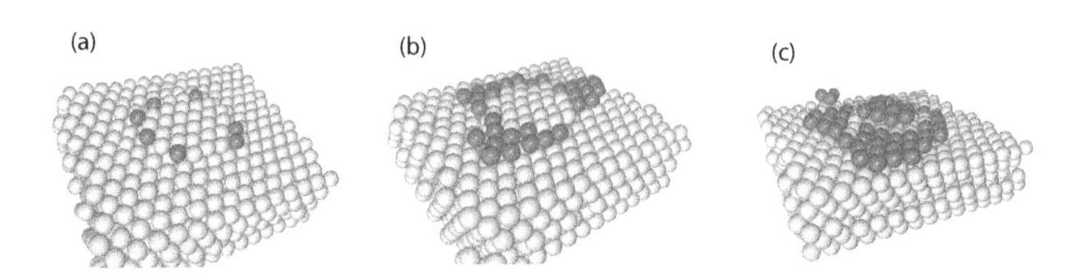

Fig. 15.7. Deposition of silver (*dark grey*) on Au(111); the potential decreases, i.e. becomes more favorable for deposition, from (a) to (c).

15.5 Growth of two-dimensional films

The phenomenon of nucleation considered is not limited to metal deposition. The same principles apply to the formation of layers of certain organic adsorbates, and the formation of oxide and similar films. We consider the kinetics of the growth of two-dimensional layers in greater detail. While the three-dimensional case is just as important, the mathematical treatment is more complicated, and the analytical results that have been obtained are based on fairly rough approximations; details can be found in [5].

A real surface of a solid metal is inhomogeneous, and nucleation for the growing clusters is favored at certain *active sites*. To simplify the mathematics we consider an electrode with unit surface area. If there are M_0 active sites, the number $M(t)$ of growing nuclei is given by first-order kinetics:

$$M(t) = M_0 \left[1 - \exp\left(-k_N t\right)\right] \tag{15.18}$$

where k_N is the rate constant for the formation of a nucleus. Two limiting cases are of particular importance:

1. $k_N t \gg 1$: instantaneous nucleation

$$M(t) = M_0 \tag{15.19}$$

which means that on the time scale considered the formation of nuclei is infinitely fast;

2. $k_N t \ll 1$: progressive nucleation

$$M(t) = k_N M_0 t \tag{15.20}$$

where at all times considered the number of nuclei is small compared to the number of active sites.

In order to derive approximate laws for the growth of a two-dimensional layer, we consider a simplified model in which all isolated clusters, i.e. clusters that do not touch another cluster, are circular. For the moment, consider a single such cluster of radius $r(t)$. New particles can only be incorporated at its boundary. Assuming that this incorporation is the rate-determining step, the number $N(t)$ of particles belonging to the cluster obeys the equation:

$$\frac{dN(t)}{dt} = 2\pi k r(t) \tag{15.21}$$

where k is the rate constant for incorporation at the boundary. This equation holds when the radius $r(t)$ is much larger than the critical radius considered in the previous section. To obtain the growth law for the radius, we express the number of particles through the area $S(t)$ covered by the cluster. If ρ denotes the number of particles per unit area, we have:

$$\frac{dS(t)}{dt} = \frac{k}{\rho} 2\pi r(t) \tag{15.22}$$

Using $S(t) = \pi r^2(t)$ a simple calculation gives:

$$r(t) = \frac{k}{\rho} t \tag{15.23}$$

Equations (15.21), (15.22), and (15.23) hold as long as the cluster does not touch the boundary of the electrode.

In a real system there will be several clusters growing simultaneously. At first the clusters are separated, but as they grow, they meet and begin to coalesce (see Fig. 15.8), which complicates the growth law. For the case of circular growth considered here, the *Avrami theorem* [6] relates the area S that is actually covered by the coalescing centers to the *extended area* S_{ex} that they would cover if they did not overlap:

$$S = 1 - \exp(-S_{\text{ex}}) \tag{15.24}$$

Note that we consider unit area; in the general case S and S_{ex} denote fractional coverage. At short times $S_{\text{ex}} \ll 1$, the clusters do not touch, and $S \approx S_{\text{ex}}$. At long times $S_{\text{ex}} \to \infty$ and $S \to 1$, and the whole surface is covered by a monolayer. For a proof of Avrami's theorem we refer to his original paper [6] (see also Problem 2).

We now consider the cases of instantaneous and progressive nucleation separately. If nucleation is instantaneous, there are M_0 growing clusters. The extended area S_{ex} is simply M_0 times the area that a single cluster would cover if it did not meet any other:

$$S_{ex}(t) = M_0 \pi r^2(t) = \frac{\pi M_0 k^2}{\rho^2} t^2 \tag{15.25}$$

So the actual area covered is:

$$S(t) = 1 - \exp\left(-\frac{\pi M_0 k^2}{\rho^2} t^2\right) \tag{15.26}$$

The concomitant current density is obtained by using $N(t) = S(t)\rho$, and:

$$j(t) = ze_0 \, \frac{dN}{dt} \tag{15.27}$$

This results in the explicit expression:

$$j = \frac{2\pi z e_0 M_0 k^2}{\rho} \, t \exp\left(-\frac{\pi M_0 k^2}{\rho^2} t^2\right) \tag{15.28}$$

for instantaneous nucleation.

In the case of progressive nucleation, new clusters are born at a constant rate $k_N M_0$. From Eq. (15.23) the area covered by a cluster born at a time t' is:

$$A(t) = \pi \frac{k^2}{\rho^2}(t - t')^2; \quad \text{for } t > t' \tag{15.29}$$

Integrating over t' and multiplying by $k_N M_0$ gives for the extended area:

$$S_{ex} = k_N M_0 \pi \frac{k^2 t^3}{3\rho^2} \tag{15.30}$$

Fig. 15.8. Overlapping circular nuclei; the extended area is the sum of the area of all the *circles* shown on the *left*.

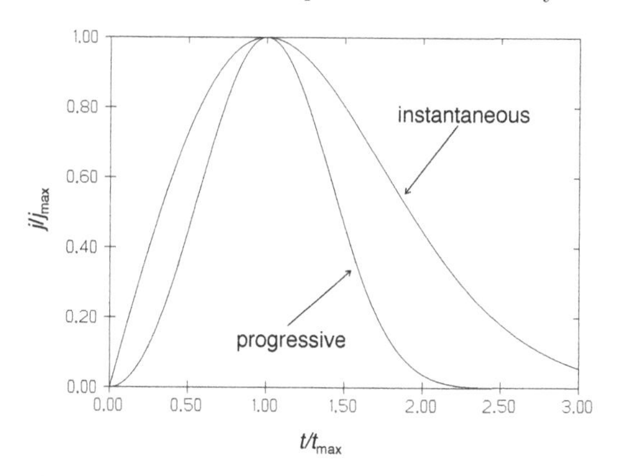

Fig. 15.9. Normalized current transients for instantaneous and progressive nucleation.

which leads to the following expression for the current density:

$$j = ze_0 k_N M_0 \pi \frac{k^2}{\rho} t^2 \exp\left(-\frac{k_N M_0 \pi k^2}{3\rho^2} t^3\right) \tag{15.31}$$

Both Eqs. (15.28) and (15.31) predict a current density which first rises as the perimeters of the clusters grow, and then decreases rapidly as the clusters begin to overlap. They can be cast into a convenient dimensionless form by introducing the maximum current density $j_{\rm max}$ and the time $t_{\rm max}$ at which it is attained. A straightforward calculation gives for instantaneous nucleation and progressive nucleation, respectivly,

$$\frac{j}{j_{\rm max}} = \frac{t}{t_{\rm max}} \exp\left(-\frac{t^2 - t_{\rm max}^2}{2t_{\rm max}^2}\right) \tag{15.32}$$

$$\frac{j}{j_{\rm max}} = \frac{t^2}{t_{\rm max}^2} \exp\left(-\frac{2(t^3 - t_{\rm max}^3)}{3t_{\rm max}^3}\right) \tag{15.33}$$

The two current transients are shown in Fig. 15.9. The curve for progressive nucleation rises faster at the beginning because not only the perimeter of the clusters increases but also their number; it drops off faster after the maximum. Such dimensionless plots are particularly useful as a diagnostic criterion to determine the growth mechanism. Real current transients may fit neither of these curves for a number of reasons, for example, if the growth starts from steps rather than from circular clusters.

15.6 Deposition on uniformly flat surfaces

Real surfaces are mostly rough and offer a multitude of growth sites. Even single crystal surfaces generally contain numerous steps and screw dislocations,

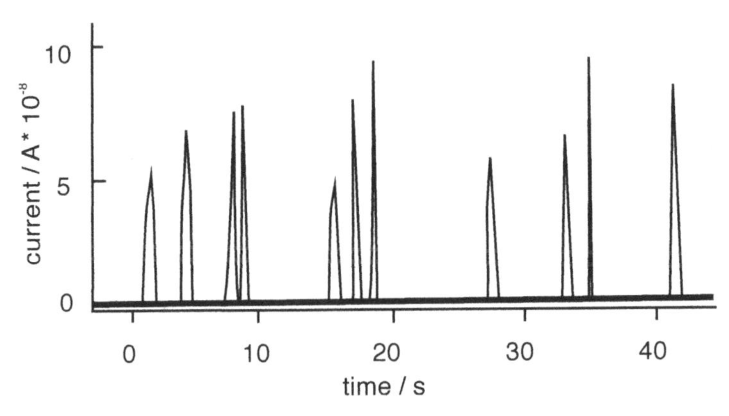

Fig. 15.10. Current pulses on a dislocation free Ag(100) surface at an overpotential of −8.5 mV. Data taken from [7].

which makes it difficult to study the deposition and growth of nuclei. However, Budewski, Kaischev and co-workers [7] have developed an elegant technique to grow flat single crystal surfaces of silver that are free of dislocations. For this purpose a suitably oriented single crystal is enclosed in a glass tube with a capillary ending. The crystal is grown further and into the capillary by slow electrolytic deposition. Any screw dislocation that is initially present will have its axis at an angle to that of the capillary tube, and hence will reach the wall as the crystal grows, and disappear from the surface. Such crystals form ideal electrodes for studying nucleation and growth phenomena. We review a few relevant experiments on dislocation-free Ag(100) surfaces in contact with a 6 M solution of $AgNO_3$.

When the electrode potential is set to a relatively low negative overpotential (of the order of 10 mV), the nucleation rate on the surface is so small that, once a nucleus has formed, it will grow into a complete monolayer before the next nucleus is formed. If the overpotential is kept constant, a series of current pulses can be observed (see Fig. 15.10), each of which corresponds to the formation and growth of a single nucleus. The integral under each peak is the charge required to form a complete monolayer of silver. The irregular spacing of the current pulses indicates that nucleation is a random event. The different heights of the spikes are due to the fact that the nuclei are formed at different sites. Nuclei that are formed nearer to the boundary of the circular electrode take longer to grow into a complete monolayer than those that are formed near the center, so that the corresponding pulses are wider and lower.

The nucleation rate k_N is the inverse of the average time between two pulses. By varying the overpotential η, the dependence of the nucleation rate on η can be obtained. In Sect. 15.3 we showed that for three-dimensional nucleation the Gibbs energy of formation is proportional to η^{-2}. A similar

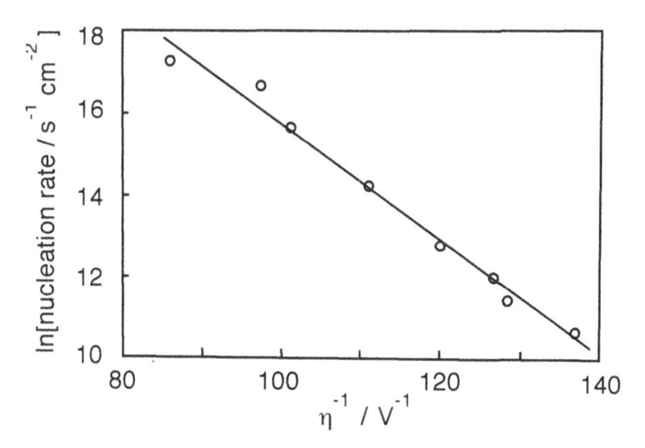

Fig. 15.11. Logarithm of the nucleation rate as a function of the inverse overpotential. Data taken from [7].

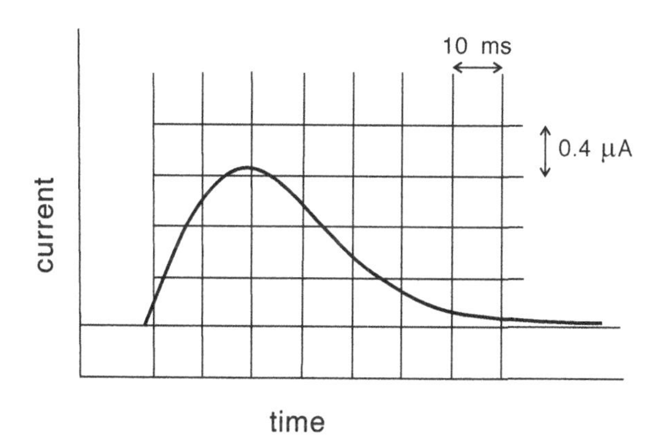

Fig. 15.12. Current transient at $\eta = -4$ mV after application of a nucleation pulse of $\eta = -17$ mV for 120 μs. Data taken from [7].

analysis for the two-dimensional case gives a proportionality to η^{-1} instead (see Problem 1). Hence a plot of $\ln k_N$ versus η^{-1} should result in a straight line, which is indeed observed (see Fig. 15.11).

The case of instantaneous nucleation can be realized by the following procedure: A sufficiently short potential pulse is applied so that a number of nuclei are formed on the surface. Subsequently the overpotential is stepped back to a low value so that existing nuclei may grow, but no new ones are formed. The resulting current transients reflect the growth of a single monolayer through (almost) instantaneous nucleation. An example can be seen in Fig. 15.12; a mathematical analysis shows that it obeys Eq. (15.28) very well.

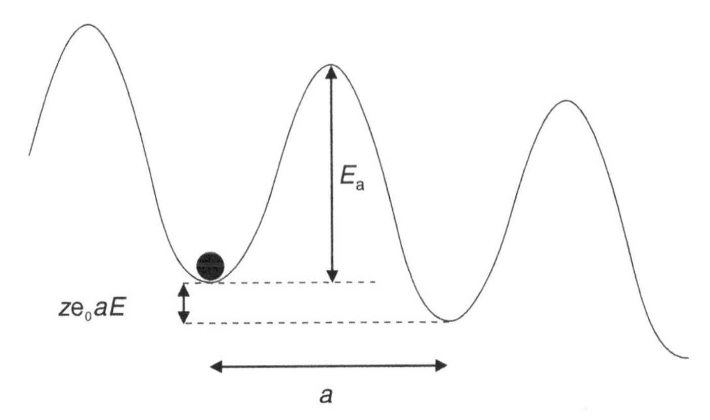

Fig. 15.13. Energy barrier for the migration of an ion in the presence of an external field.

When a high (negative) overpotential is applied a second layer can begin to grow before the first one is completed. This leads to multilayer growth, which is only imperfectly understood, so we refrain from a further discussion.

15.7 Metal dissolution and passivation

Metal dissolution is the inverse process to the deposition; so its principles can be derived from preceding considerations. It should, however, be borne in mind that the preferred sites for deposition need not be the same as those for the dissolution. This is particularly true if the reactions are far from equilibrium. Therefore, rapid cycling of the potential between the deposition and the dissolution region can lead to a substantial roughening of the electrode surface, which can be used in techniques such as surface-enhanced Raman spectroscopy, which require a large surface area.

Often the dissolution of a metal leads to the formation of an oxide film on the electrode surface. These films are usually nonconducting and hinder the further dissolution of the metal, a phenomenon known as *passivation*. Such passive-film formation is prevalent with the valve metals such as aluminum and titanium. In aqueous solutions aluminum forms an oxide film according to the reaction:

$$2\mathrm{Al} + 3\mathrm{H_2O} \rightleftharpoons \mathrm{Al_2O_3} + 6\mathrm{H^+} + 6\mathrm{e^-} \tag{15.34}$$

The resulting films can attain thicknesses of the order of a thousand Ångstroms or more.

Once the film has began to form, ions must pass through the film in order for the reaction to proceed. The general case is quite complicated since the films can have both an ionic and an electronic conductivity. We consider the simple case of an electronically insulating, homogeneous film, and assume that one kind of ion, with charge number z, can migrate in the presence of an external field E. The migrating ion can occupy certain sites within the film, and migration consist of a series of thermally activated jumps between these sites. Let a denote the distance between adjacent sites, and E_a^0 the energy of activation for a jump in the absence of an applied field (see Fig. 15.13). Application of a field E creates a drop aE in the electrostatic potential between adjacent sites, and an energy gain ze_0aE per jump (to be specific, we assume $z > 0$). This entails a change in the energy of activation, a phenomenon known from electron-transfer reactions (see Chap. 9, in particular Fig. 9.1). If the barrier is symmetric, the energy of activation will be lowered by an amount $ze_0aE/2$, which corresponds to a transfer coefficient of 1/2. The rate of jumps in the forward direction is then:

$$k_f = \nu \ \exp \frac{E_a^0 - ze_0aE/2}{kT} \tag{15.35}$$

where ν is the frequency factor. Similarly the energy of activation for the backward direction increases by $ze_0aE/2$, so the backward rate is:

$$k_b = \nu \ \exp\left(-\frac{E_a^0 + ze_0aE/2}{kT}\right) \tag{15.36}$$

The concomitant current density is:

$$j = ze_0n(k_f - k_b) = 2ze_0n\nu \ \exp\left(-\frac{E_a^0}{kT} \sinh \frac{ze_0aE}{2kT}\right) \tag{15.37}$$

where n is the density of ions. For small fields the formula can be linearized:

$$j = ze_0n\nu \ \exp\left(-\frac{E_a^0}{kT}\right) \frac{ze_0aE}{kT}, \qquad \text{for } ze_0aE \ll kT \tag{15.38}$$

while for large fields the back current can be neglected, and the current depends exponentially on the field:

$$j = ze_0n\nu \ \exp\left(-\frac{E_a^0}{kT}\right) \ \exp \frac{ze_0aE}{2kT}, \qquad \text{for } ze_0aE \gg kT \tag{15.39}$$

Note that the field is the important variable, not the electrode potential. Typically fields of the order of 10^6 V cm^{-1} are required to produce a noticeable film growth.

The growth law of Eq. (15.37) is often observed on valve metals. From the growth at high fields the average jump distance a can be calculated. Steady-state measurements give surprisingly large values of the order of 5 Å or even higher [8, 9]. In contrast pulse measurements give smaller values of the order of 2 Å, which fit better into the microscopic model on which Eq. (15.37) is based. An external field induces structural changes in the oxide film; in particular, high fields seem to produce pairs of vacancies and interstitials which enhance the current. A model that accounts for these changes leads to an equation of the same form as Eq. (15.37) [10], but a no longer has the meaning of a jump distance. Therefore, only the values obtained from pulse measurements provide an estimate of the jump distance, while the formal parameter a obtained from steady-state measurements has no direct interpretation.

Problems

1. Consider the formation of a two-dimensional nucleus and show that the Gibbs energy of a critical cluster is inversely proportional to η. For this purpose introduce a boundary energy which is proportional to the perimeter of the cluster.
2. Here we derive Avrami's theorem for a simple case [6]. Consider an area A that is partially covered by N circles each of area a, where $a \ll A$. The circles overlap so that the area that is actually covered is smaller than the extended area Na. Show that the probability that a particular point is not covered by any circle is:
$$\left(1 - \frac{a}{A}\right)^N = \left(1 - \frac{na}{N}\right)^N \tag{15.40}$$
where $n = N/A$ is the density of clusters. Using the theorem:
$$\lim_{N\to\infty} \left(1 - \frac{x}{N}\right)^N = e^{-x} \tag{15.41}$$
show that in the limit of infinitely many clusters the probability that a point is not covered is given by:
$$\exp(-na) = \exp\left(-S_{\text{ex}}\right) \tag{15.42}$$
Hence the fraction of the surface that is covered is:
$$S = 1 - \exp\left(-S_{\text{ex}}\right) \tag{15.43}$$
Note that S and S_{ex} relate to unit area. The generalization of this argument to randomly placed clusters of arbitrary size and shape is given in [6].
3. Consider the formation of hemispherical nuclei of mercury on a graphite electrode. The interfacial tension of mercury with aqueous solutions is about 426 mN m^{-1}. From Eq. (15.16) calculate the critical cluster sizes for $\eta = -10, -100, -200$ mV. Take $z = 1$ and ignore the interaction energy of the base of the hemisphere with the substrate.

References

1. W.R. Fawcett, *Langmuir* **5** (1989) 661
2. F. van der Pool, M. Sluyters-Rehbach, and J.H. Sluyters, *J. Electroanal. Chem.* **58** (1975) 177; R. Andreu, M. Sluyters-Rehbach, and J.H. Sluyters, *J. Electroanal. Chem.* **134** (1982) 101; ibid. 171 (1984) 139.

3. E. Budevski, G. Staikov, und W.J. Lorenz, *Electrochemical Phase Formation and Growth.* VCH, Weinheim, 1996 (A good general source on this topic is the book).
4. M. Mariscal, E. Leiva, K. Pötting, and W. Schmickler, *Appl. Phys. A* **87** (2007) 385.
5. E. Bosco and S.K. Rangarajan, *J. Electroanal. Chem.* **134** (1981) 213.
6. M. Avrami, *J. Chem. Phys.* **7** (1937) 1130; **8** (1940) 212; **9** (1941) 177.
7. R. de Levie, *Advances in Electrochemistry and Electrochemical Engineering*, Vol. 13, edited by H. Gerischer and W. Tobias. Wiley Interscience, New York, NY, 1985.
8. E.B. Budewski, *Progress in Surface and Membrane Science*, Vol. 11, edited by D.A. Cadenhead and J.F. Daniell. Academic Press, London, 1976; E.B. Budewski, *Treatise of Electrochemistry*, Vol. 7, edited by J. O'M. Bockris, B.E. Conway, and E. Yeager. Plenum Press, New York, 1983.
9. D.A. Vermilyea, *Advances in Electrochemistry and Electrochemical Engineering*, Vol. 3, edited by P. Delahay and W.C. Tobias. Wiley Interscience, New York, NY, 1963.
10. L. Young, *Anodic Oxide Films.* Academic Press, London, 1961.
11. J.F. Dewald, *J. Electrochem. Soc.* **104** (1957) 244.

16

Electrochemical surface processes

Usually electrochemical processes involve an exchange of electrons or ions between the electrode and the electrolyte. The driving force is the difference in the electrochemical potential of the transferring species, and therefore depends on the electrode potential. However, there exist processes which take place on the electrode surface only, without exchange of charge with the solution. Nevertheless, these processes often show a strong dependence on the electrode potential. At a first glance, this may seem strange: the electrode surface is equipotential, and hence a change of the potential should not affect the driving force directly. However, a change of potential entails a change of the surface charge density, and hence of the electric field in the double layer. This field interacts with local dipole moments on the surface, and may thereby affect surface processes. In this chapter, we will discuss a few examples. A good general reference to surface processes in uhv and in electrochemistry is the book by Ibach [1].

Besides this field-dipole coupling, there is a second mechanisms which may also affect surface processes. A change in the electrode potential often involves a change in the coverage of adsorbates, particularly of adsorbed anions, and this will in turn influences surface processes. An example is the adsorption of chloride ions, which greatly enhances the mobility of gold atoms. Since little is known about the details of this mechanism, we shall not consider it further.

In addition, we treat the topic of *self-assembled layers*, which form spontaneously on electrode surfaces under certain conditions, though at the time of writing we cannot exclude that some of them might entail charge exchange.

16.1 Surface reconstruction

In the vacuum, several single crystal surfaces are reconstructed. In these cases, the perfectly terminated bulk surfaces are not stable, but the surface atoms rearrange to form a denser, energetically more favorable structure. A good

W. Schmickler, E. Santos, *Interfacial Electrochemistry*, 2nd ed.,
DOI 10.1007/978-3-642-04937-8_16, © Springer-Verlag Berlin Heidelberg 2010

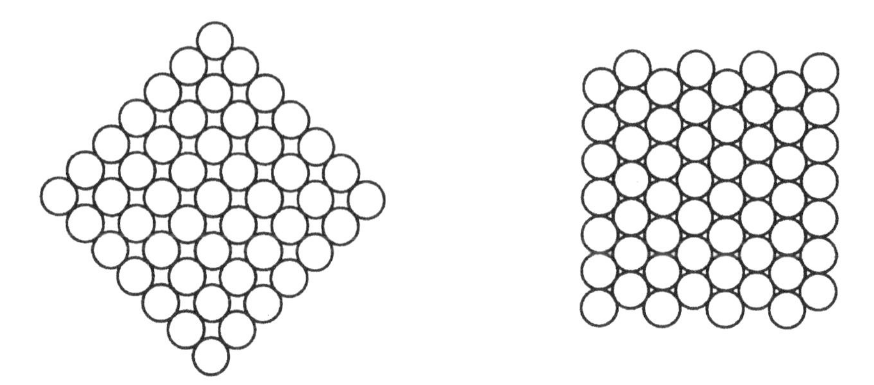

Fig. 16.1. Perfect (*left*) and reconstructed (*right*) Au(100) surface (schematic).

example is the Au(100) surface, which in the vacuum forms a reconstructed surface with hexagonal structure. The ideal structures are shown in Fig. 16.1. Since the hexagonal structure is denser, there is a mismatch with the underlying layer; so a corrugated surface with a local hexagonal structure is formed – for clarity, this corrugation is not shown in the figure. In aqueous solutions, the reconstructed surface is stable at low potentials, but at higher potentials it is lifted, and the perfectly terminated surface reappears. The potential, at which this lifting occurs, depends on the composition of the electrolyte; in weakly adsorbing solutions like perchloric acid, it occurs at about 0.55 V vs. SCE.

When the potential is stepped back below the transition point, the reconstructed surface reappears. On Au(100) both the lifting and the formation of the reconstruction are slow, so that the capacities of both surfaces can be obtained over a fairly large potential range, including regions in which they are not thermodynamically stable (see Fig. 16.2). This makes it possible to elucidate the thermodynamics of this process in some detail.

As pointed out in Sect. 4.4, the correct thermodynamic function for an electrode held at constant potential is the surface tension γ, which is a function of the electrode potential ϕ. The reconstructed and the unreconstructed surfaces have different surface tensions, and at each potential the surface with the lower surface tension is the one that is thermodynamically stable. Therefore, to understand the driving force we require the surface tension of both modifications as a function of potential. We focus on the basic case of a non- or weakly adsorbing electrolyte.

From the capacity minima that occur at sufficiently dilute (i.e.$\leqslant 10^{-2}$ M, see Chap. 5) solutions, we determine the potential of zero charge (pzc) ϕ_0 of both surfaces. The reconstructed surface has the higher work function and hence the higher pzc. By integrating the capacity, we obtain the charge density σ as a function of the potential. Integrating again (see Eq. 4.13) gives the

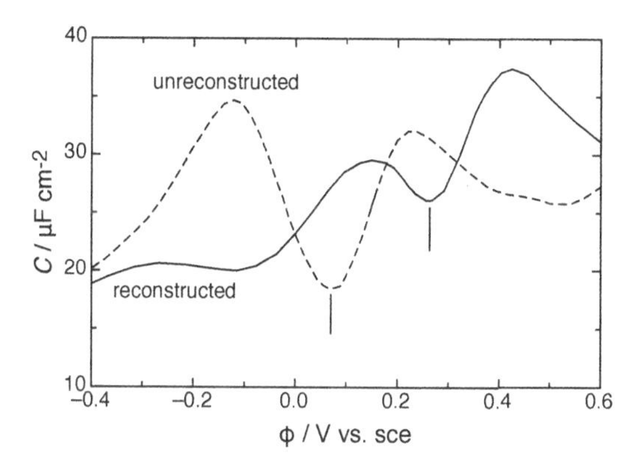

Fig. 16.2. Capacity curves for the reconstructed and the bulk terminated Au(100) surface in 10 mM perchloric acid. The potentials of zero charge are indicated by *vertical lines*. Data taken from [2].

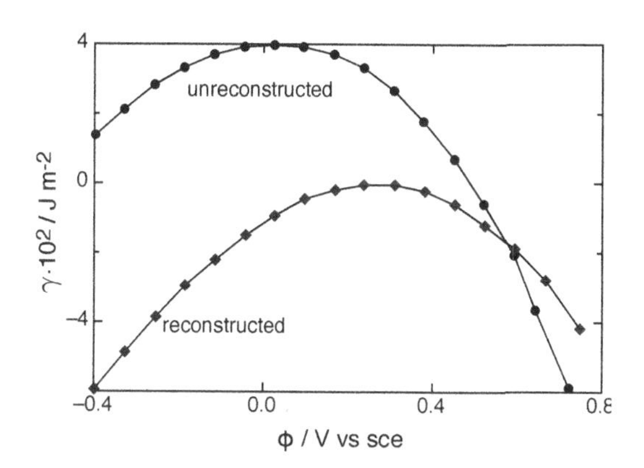

Fig. 16.3. Surface tension for the reconstructed and the bulk terminated Au(100) surface in 10 mM perchloric acid. Data taken from [3]

surface tension:

$$\gamma(\phi) = \gamma_0 - \int_{\phi_0}^{\phi} \sigma(\phi')\, d\phi' \tag{16.1}$$

up to the unknown value γ_0 at the pzc. Of course, this equation holds for both surfaces.

While it is not possible to determine the absolute values of the surface tension at the pzc, we can obtain the difference $\Delta\gamma_0 = \gamma_0^{\text{unrecon}} - \gamma_0^{\text{recon}}$. For this purpose, we arbitrarily set $\gamma_0^{\text{recon}} = 0$ and choose $\gamma_0^{\text{unrecon}}$ such that the two surface tension curves intersect at the potential, at which the lifting of the reconstruction occurs experimentally (near 0.55 V vs. SCE). The resulting curves, shown in Fig. 16.3, provide a thermodynamic description of the reconstruction and its lifting. At potentials below the crossing point, the reconstructed surface has the lower surface tension and is thermodynamically stable, at higher potentials it is the unreconstructed. This construction also provides an estimate for the change $\Delta\gamma_0$ in surface tension during the reconstruction of the uncharged surface: $\Delta\gamma_0 = (4.1 \pm 0.3) \times 10^{-2}$ Jm^{-2}. This is the value for a surface immersed in aqueous solution, but since the interaction of gold with water is weak, the value for the same surface in vacuum should be close.

This procedure can be applied whenever the capacity curves for both surfaces can be measured over a sufficiently wide range, and is also valid in the presence of specific adsorption. Another example is Au(111), which in the vacuum also reconstructs. In a solution of non-adsorbing electrolyte, the reconstruction is lifted near 0.4 V vs. SCE. Since the bulk terminated surface of Au(111) is densely packed to start with, the gain in energy during reconstruction is much smaller than on Au(100): In this case, the best estimate is $\Delta\gamma_0 = (3 - 5) \times 10^{-3}$ Jm^{-2}.

The lifting of the reconstruction on these two gold surfaces can be understood in terms of the field-dipole interaction mentioned above. In both cases, the reconstructed surfaces have a higher work function; this implies that they have a larger surface dipole. The surface dipole moment μ is always directed towards the bulk, and it interacts with the electric field E_0 in the double layer with an energy $-\mu E_0$. Here, E_0 is the unscreened field. At potentials above the pzc, this interaction energy is positive and is the larger, the greater the surface dipole. Hence with increasing potential the reconstructed surface becomes less favorable.

16.2 Steps, line tension and step bunching

Steps and islands are common features on electrode surfaces, and their energetics and dynamics are interesting topics in their own right. In a certain sense, steps are the one-dimensional analogues of surfaces. Thus, in analogy with the surface tension, we define the step line tension β as the extra energy caused by the presence of a step. More precisely, it is the extra surface tension caused by the step, since the latter is the correct thermodynamic energy. Just like the surface tension, the line tension depends strongly on the electrode potential, but here the analogy ends, because this dependence is quite different, as we shall demonstrate.

Every step has an associated dipole moment, which is caused by the Smoluchowski [4] effect. At the step edge, the positive charge residing at the atom cores drops abruptly, while the electronic density changes smoothly. As can be seen from Fig. 16.4, this leads to a positive excess charge at the outer step edges, which is balanced by a negative charge near the foot. This results in a dipole moment pointing towards the solution. It is convenient to measure

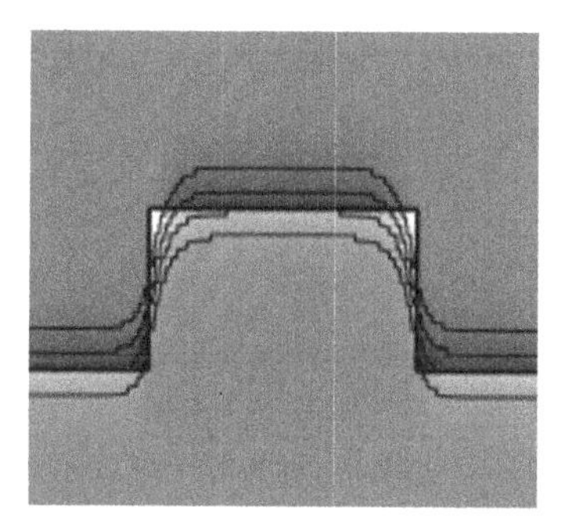

Fig. 16.4. Charge distribution at a step; the *dark regions* denote an excess of negative charge, the *white regions* of positive charge.

the step dipole μ per atom in terms of the unit charge times Ångstroms, and typical values are of the order of 10^{-2} e_0Å. Since the dipole is directed outwards, it entails a reduction of the work function. For a vicinal surface, which has a uniform step density, the change in the work function compared with the perfectly smooth surface is determined by the average dipole moment per area:

$$\Delta\Phi = -\frac{\mu}{\epsilon_0 a_{\|} L} \tag{16.2}$$

where L is the width of the terraces between the steps, and $a_{\|}$ the step length per atom. In the absence of specific adsorption, the same relation should hold for the potential of zero charge. In the few cases were this has been tested, it was indeed fulfilled; an example is shown in Fig. 16.5. From the slope of the plot the step dipole moment can be determined. Values obtained for the step dipole in aqueous solutions are usually close to those obtained in uhv. This indicates that the presence of water does not greatly affect the local dipole moment [5]

The dipole moment of the steps interacts with the electric field $E = \sigma/\epsilon_0$ produced by the charge density σ. This interaction dominates the dependence of the step line tension on the potential or on the charge [6]. Therefore:

$$\beta = \beta(\phi_0) - \frac{\mu}{\epsilon_0 a_{\|}}\sigma \tag{16.3}$$

There are correction terms caused by the polarizability of the step dipole, and by the double layer structure at the steps. However, in the absence of specific adsorption, and in the vicinity of the pzc, this is a good approximation. Just like the surface tension, the absolute value of the line tension cannot be measured by electrochemical techniques, but the variation can be obtained from the difference between the surface tension of a stepped surface and a flat surface. An example is shown in Fig. 16.6; in this case relation 16.3 is well obeyed. At larger positive charge densities, there is always some specific adsorption of anions, and major deviations are observed [7].

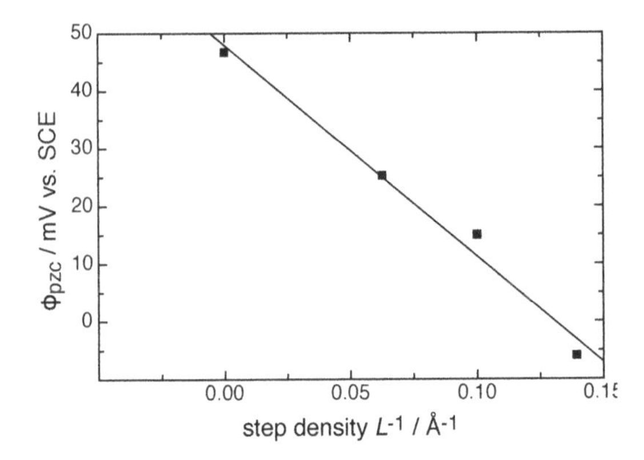

Fig. 16.5. Potential of zero charge versus the density of steps on a Au(100) surface. Data taken from [7].

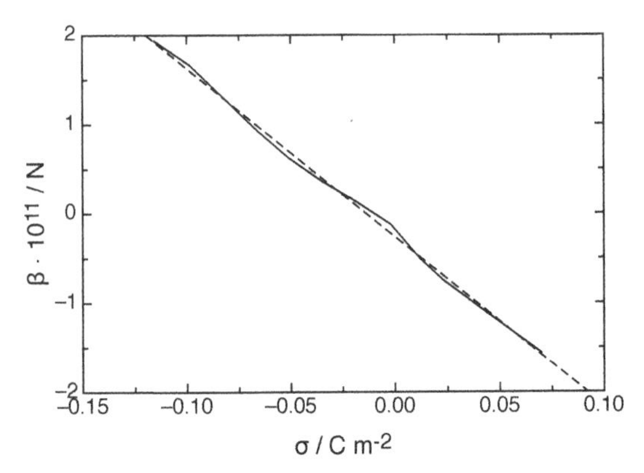

Fig. 16.6. Variation of the step line tension with charge density. Data taken from [7], plot courtesy of G. Beltramo, Forschungszentrum Jülich. The *full line* are the experimental data, the *dotted line* is a fit to Eq. (16.3).

An interesting consequence of Eq. (16.3) is the possibility, that the step line tension may vanish at sufficiently positive charge densities. In this case the steps would become unstable and dissolve. This effect, which has not yet been observed in electrochemistry, would be the analogue of a *roughening transition* [1] observed in uhv.

The presence of steps induces a stress on the surface, and therefore on a bare surface in uhv the steps repel each other, and vicinal surfaces are stable. In contrast, on stepped electrodes there is a thermodynamic driving force for step bunching. A nice example is shown in Fig. 16.7, where the fairly regular steps on a Ag(19 19 17) surface separate with time into a Ag(111) terrace and a part in where the steps are bunched.

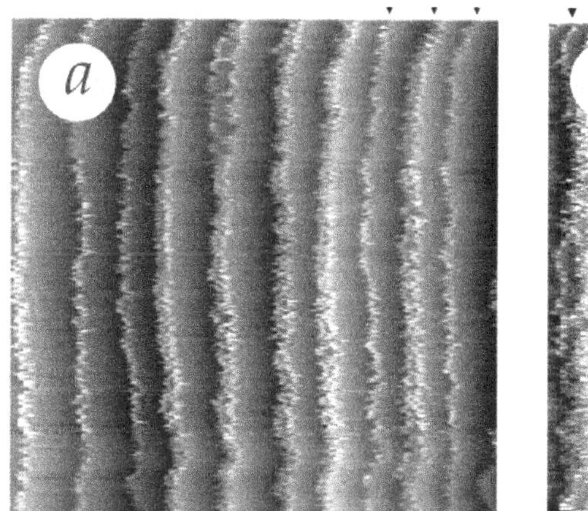

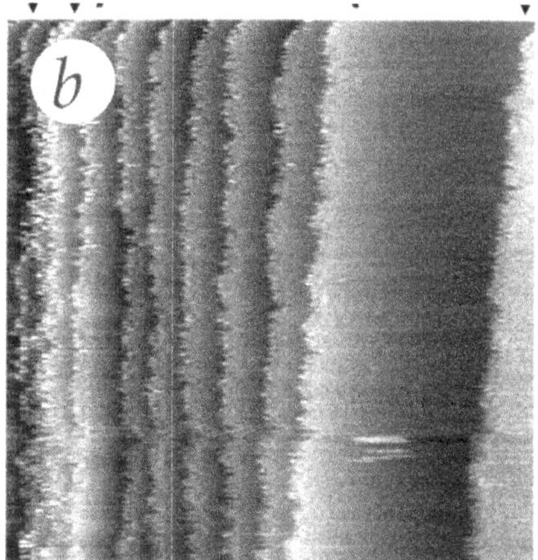

Fig. 16.7. Freshly prepared vicinal Ag(19 19 17) surface (**a**) and the same surface about 40 min later (**b**); after [8]; scan range: 100 nm.

The different behavior in uhv and in electrolyte solutions is caused by the boundary conditions. As pointed out repeatedly, in electrochemistry the correct thermodynamic energy is the surface tension. This has a maximum at the pzc, and drops off, roughly quadratically, on both sides. Hence, it is energetically favorable for an electrode to be far away from the pzc. However, since the potential is held constant, the only way the electrode can move away from the pzc is by changing its surface structure. This is what happens during surface reconstructions. On a stepped surface, step bunching divides the surface into two parts with different surface tension, such that the total is lower than that for the regularly stepped surface. The details can be found in [9].

16.3 Surface mobility

There are many instances, in which the surface mobility increases with the potential. An example is the diffusion of gold atoms on a gold electrode shown in Fig. 16.8. The underlying mechanism is again the interaction of a local dipole moment with the double-layer field.

Just like a step, a single metal atom on a flat terrace generally has a dipole moment with the positive end pointing towards the solution, caused by the Smoluchowski effect (see Fig. 16.9). The magnitude of the dipole depends on the position: It is lowest in the equilibrium position, which is a hollow site, and larger at a bridge site, because the distance from the surface is larger. The bridge sites form the barrier for the migration of the atom. Let us denote the dipole moment in the equilibrium position by μ_i, and at the barrier by

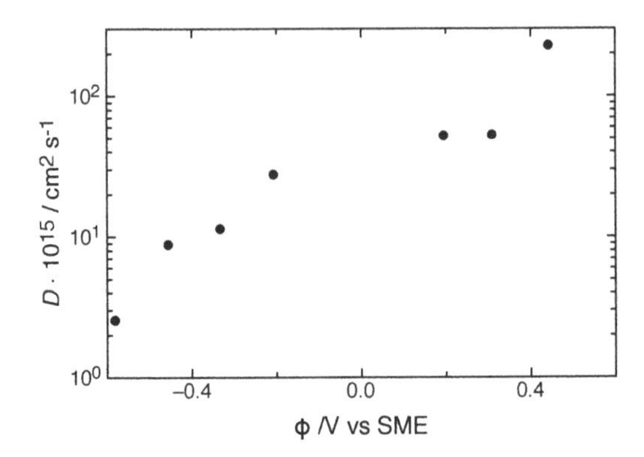

Fig. 16.8. Apparent diffusion coefficient for Au atoms on a gold electrode. Data taken from [10].

$\mu^\dagger$. In both positions the dipole interacts with the field $E = \sigma/\epsilon_0$. Therefore, the difference in the interaction energy enters into the energy of activation, and we obtain for the surface diffusion coefficient the relation:

$$D \propto \exp -\frac{(\mu^\dagger - \mu_i)E}{kT} \tag{16.4}$$

and the rate of migration increases exponentially with the field. In addition,

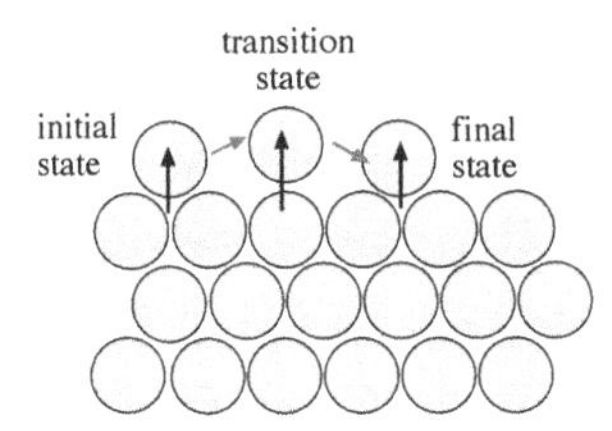

Fig. 16.9. Change in the dipole moment during adatom migration.

the concentration of migrating adatoms also increases with the field, since the energy of the adatom contains a term $-\mu_i E$, which also enters exponentially into the equilibrium concentration.

The increasing mobility affects a process known as *Ostwald ripening.* When an electrode surface contains metal islands of various sizes, the larger islands grow at the expense of the smaller ones, because larger islands have a more favorable ratio of bulk energy to boundary energy (see Problem 1). Figure 16.10 shows a series of STM images showing the gradual disappearance of a small island next to a bigger one. Since the ripening occurs via adatom migration, it can be enhanced by the field-dipole interaction [1, 9].

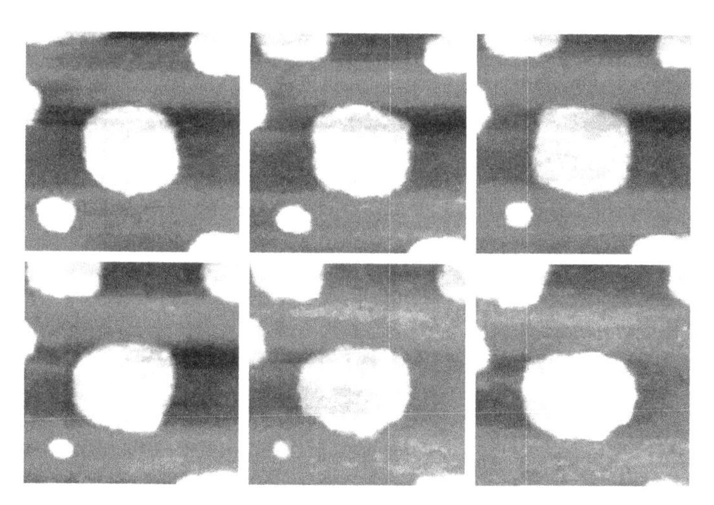

Fig. 16.10. Ostwald ripening of islands on Au(111). Note that the small island in the *lower left corner* disappears gradually. Courtesy of M. Giesen, Forschungszentrum Jülich.

16.4 Self-assembled monolayers (SAMs) in electrochemistry

The concept of self-assembling is astonishing, and especially in biological systems it is ubiquitous. The best example is our brain, which is an intricate ensemble of neurons grouped into modules without a command centre; all regions are connected by multiple bidirectional pathways, making the brain precisely the paradigm of a self-organizing distributed system.

The formation of monolayers by self-assembling of surfactant molecules at surfaces is a simpler example of the general phenomena of self-assembly. The molecules that form SAMs consist of three parts (see Fig. 16.11): a headgroup that binds to the surface, an organic moiety (in the most simple case an alkyl chain), and a terminal functional group which interacts with the environment. The packing and ordering of the layer result from a balance and interplay of various forces. The adsorbates adopt a geometric arrangement that minimizes the free energy of the layer and allows a high degree of van der Waals, electrostatic, and steric interactions, and in some cases hydrogen bonds with the neighboring molecules are formed. Since the *free* energy is minimized, entropic effects also contribute to the final conformation.

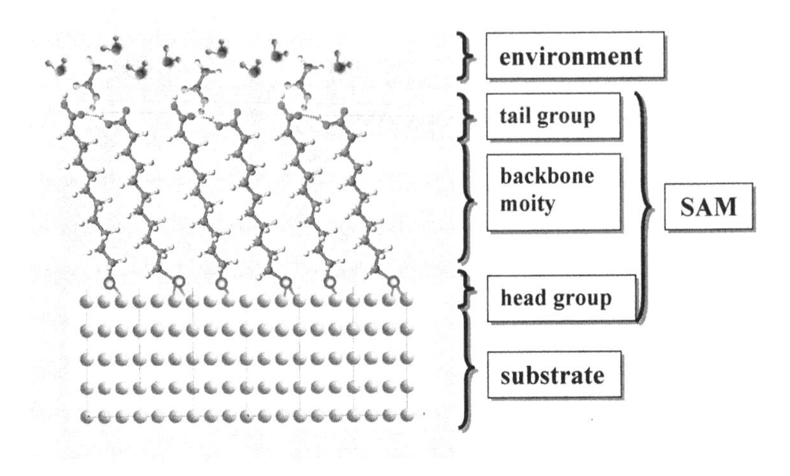

Fig. 16.11. Structure of self-assembled monolayer.

The self-assembling process can occur on various substrate surfaces. There are specific linkers for each type of substrate. Head groups containing sulfur or nitrogen are appropriate for clean metals, silicon and phosphor for hydroxilated and oxidized surfaces. The most extensively investigated SAMs are alkanethiols on gold and silver, but they can also be formed on semiconductor surfaces such as SiO_2 and GaAs. Considering the bonding arrangement formed at the metal – sulfur interfaces, the molecules comprising the SAM tend to adopt structural arrangements that are similar to simple adlayer structures formed by elemental sulfur on that metal. Thus, the generally accepted structures of thiols on Au(111) at high coverages is $(\sqrt{3} \times \sqrt{3})R30°$, and on Ag(111) it is $(\sqrt{7} \times \sqrt{7})R10.9°$, like the overlayers resulting from the adsorption of SH_2 or sulfide salts. However, also adsorbates on Au (111) with two non-equivalent chains alternating their orientations have been proposed to exist in a unit cell defining a $c(4 \times 2)$ superlattice structure. The specific ordering of the sulfur determines the free space available to the organic moiety. The alkyl chains organize themselves within the constraints imposed by the structure of the adlayer. However, steric crowding of bulky substituents in the alkyl chain can determine a less dense packing structure of the sulfur arrangement. The metal-sulfur bonding drives the structural configuration of the adlayer and determines the maximal coverage, while the attractive lateral interactions between the organic moieties promote the secondary organization of the alkyl chains. Each methyl group contributes about 1 kcal mol^{-1} to the stabilization of the SAM. The alkyl chains adopt a quasi-crystalline structure, where the chains are fully extended in a nearly all-trans conformation. The tilt angle of the backbone chain is about 30° for SAMs on Au, while on silver is mostly highly oriented along the surface normal direction (10°).

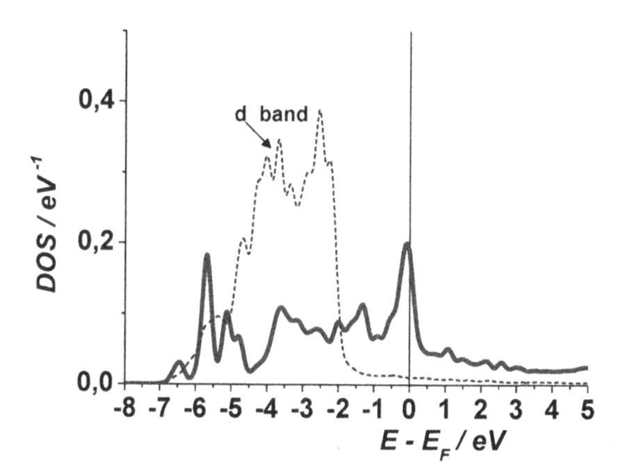

Fig. 16.12. Densities of state of the d band of Au(111) and of the s and p orbitals of sulphur for SAMs containing propanthiol.

The mechanism of the bond formation between the metal and the sulfur atom is still controversial. It is not clear, if at all and how, the S–H bond is broken. Where does the hydrogen go? Neither gold nor silver are metals that strongly adsorb hydrogen. It seems probable that SAM formation in vacuum leads to a loss of the hydrogen in the form of H_2 molecules. Also the nature of the bond, when a thiolate species (R-$S^-M^+ \cdot M_n^\circ$) results or a covalent bond is formed, is under discussion. Figure 16.12 shows the results of DFT calculations for the interaction of the $s-p$ orbitals of sulfur in the propanethiol radical with the d band of Au(111). The coupling is very strong, as can be observed by the broading of the orbitals (compare with Figs. 14.4 and 14.5).

SAMs can link the external environment to the electronic and optical properties of metallic surfaces. They act as nanostructures themselves with well-defined shapes and sizes, and form patterns on surfaces with critical dimensions below 100 nm and thicknesses of the order of 1–3 nm. The composition of the tail groups determines the properties of the interface and the interaction with the environment. They can also be formed on other nanosystems (nanoparticles, for instance), and they can specifically interact with biological nanostructures such as proteins.

The recent accelerated development in nano-science has given a new impulse to the topic of self-assembled monolayers. The early ideas of the 1980s, to build nanodevices with SAMs, now appear as a real possibility. However, before a SAM-based nanotechnology becomes real, a number of obstacles have to be overcome. For an extensive discussion of SAMs we refer to a number of excellent books and articles [11–14].

Problems

1. Consider two circular metal islands, one layer of atoms high, consisting of the same material. Let R_1 and R_2, with $R_1 > R_2$, denote their radii, and β their step line tension. Calculate the gain in free energy, when the larger island completely swallows the smaller one. Consider in particular the case $R_1 \gg R_2$.
2. We consider the migration of a single adatom on the surface of a metal electrode. Let the difference in dipole moment between the equilibrium position and the activated state be: $(\mu^\dagger - \mu_i)) = 5 \times 10^{-3} e_0$Å. Calculate the enhancement in the migration rate caused by surface-charge densities of 5, 10, 20 μCcm^{-2}.

References

1. H. Ibach, *Physics of Surfaces and Interfaces*, Springer, Berlin, Heidelberg, 2006.
2. D. Eberhardt, Diploma Thesis, University of Ulm, Germany, 1995.
3. E. Santos and W. Schmickler, *Chem. Phys. Lett.* **400** (2005) 26.
4. R. Smoluchowski, *Phys. Rev.* **60** (1941) 661.
5. G. Beltramo, H. Ibach, and M. Giesen, *Surf. Sci.* **601** (2007) 1876.
6. J. Lecoeur, J. Andro, and R. Parsons, *Surf. Sci.* **114** (1982) 326.
7. H. Ibach and W. Schmickler, *Phys. Rev. Lett.* **91** (2003) 016106.
8. A. Hamelin and J. Lecoeur, *Surf. Sci.* **57** (1976) 771.
9. S. Baier, H. Ibach, and M. Giesen, *Surf. Sci.* **573** (2004) 17.
10. M. Giesen, H. Ibach, and W. Schmickler, *Surf. Sci.* **573** (2004) 24.
11. J. Gonzales Velasco, *Chem. Phys. Lett.* **312** (1997) 7.
12. A. Ulman, *Chem. Rev.* **96** (1996) 1533.
13. J.C. Love, L.A. Estroff, J.K. Kriebel, R.G. Nuzzo, and G.M. Whitesides, *Chem. Rev.* **105** (2005) 1103.
14. C. Vericat, M.E. Vela, G.A. Benitez, J.A.M. Gago, X. Torreles, and R.C. Salvarezza, *J. Phys. Condens. Matter* **18** (2006) R867.
15. U. Ulman, *An Introduction to Ultrathin Organic Films: from Langmuir-Blodgett to Self-Assembly*. Academic Press, San Diego, CA, 1991.

17

Complex reactions

In previous chapters we have already encountered examples of reactions involving several steps, and introduced the notion of a *rate-determining step*. Here we will elaborate on the subject of complex reactions, introduce another concept; the *electrochemical reaction order*, and consider a few other examples.

17.1 Consecutive charge-transfer reactions

The simplest type of complex electrochemical reactions consists of two steps, at least one of which must be a charge-transfer reaction. We consider two consecutive electron-transfer reactions of the type:

$$\mathrm{Red} \rightleftharpoons \mathrm{Int} + \mathrm{e}^- \rightleftharpoons \mathrm{Ox} + 2\mathrm{e}^- \tag{17.1}$$

such as:

$$\mathrm{Tl}^+ \rightleftharpoons \mathrm{Tl}^{2+} + \mathrm{e}^- \rightleftharpoons \mathrm{Tl}^{3+} + 2\mathrm{e}^- \tag{17.2}$$

For simplicity we assume that the intermediate is short-lived, stays at the electrode surface, and does not diffuse to the bulk of the solution. Let $\phi_{00}^{(1)}$ and $\phi_{00}^{(2)}$ denote the standard equilibrium potentials of the two individual steps, and c_{red}, c_{int}, c_{ox} the surface concentrations of the three species involved. If the two steps obey the Butler–Volmer equation the current densities j_1 and j_2 associated with the two steps are:

$$\begin{aligned} j_1 = F k_1^0 \Bigg[& c_{\mathrm{red}} \exp \frac{\alpha_1 F(\phi - \phi_{00}^{(1)})}{RT} \\ & - c_{\mathrm{int}} \exp \left(- \frac{(1-\alpha_1) F (\phi - \phi_{00}^{(1)})}{RT} \right) \Bigg] \end{aligned} \tag{17.3}$$

W. Schmickler, E. Santos, *Interfacial Electrochemistry*, 2nd ed.,
DOI 10.1007/978-3-642-04937-8_17, © Springer-Verlag Berlin Heidelberg 2010

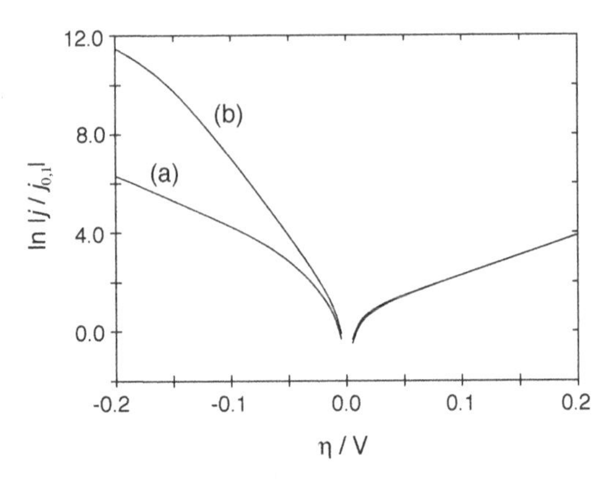

Fig. 17.1. Tafel plot for two consecutive electron-transfer reactions. Parameters: $\alpha_1 = 0.4$, $\alpha_2 = 0.5$; (a) $j_{0,2} = 5 j_{0,1}$; (b) $j_{0,2} = 10^3 j_{0,1}$.

$$
\begin{aligned}
j_2 = F k_2^0 \Bigg[& c_{\text{int}} \exp \frac{\alpha_2 F(\phi - \phi_{00}^{(2)})}{RT} \\
& - c_{\text{ox}} \exp \left(- \frac{(1-\alpha_2) F(\phi - \phi_{00}^{(2)})}{RT} \right) \Bigg]
\end{aligned}
\tag{17.4}
$$

The total current density is $j = j_1 + j_2$. Let us first consider the equilibrium conditions. From $j_1(\phi_0) = j_2(\phi_0) = 0$ we obtain:

$$
\phi_0 - \phi_{00}^{(1)} = \frac{RT}{F} \ln \frac{c_{\text{int}}(\phi_0)}{c_{\text{red}}} \tag{17.5}
$$

$$
\phi_0 - \phi_{00}^{(2)} = \frac{RT}{F} \ln \frac{c_{\text{ox}}}{c_{\text{int}}(\phi_0)} \tag{17.6}
$$

from which the equilibrium potential ϕ_0 and the concomitant concentration $c_{\text{int}}(\phi_0)$ can be determined:

$$
c_{\text{int}}(\phi_0) = (c_{\text{ox}} c_{\text{red}})^{1/2} \exp \left(- \frac{F\left(\phi_{00}^{(1)} - \phi_{00}^{(2)}\right)}{2RT} \right) \tag{17.7}
$$

$$
\phi_0 = \frac{\phi_{00}^{(1)} + \phi_{00}^{(2)}}{2} + \frac{RT}{2F} \ln \frac{c_{\text{ox}}}{c_{\text{red}}} \tag{17.8}
$$

On application of an overpotential η we have under stationary conditions:

$$
j(\eta) = 2 j_1(\eta) = 2 j_2(\eta) \tag{17.9}
$$

Substituting from above gives:

$$j_1(\eta) = j_{0,1} \left[\exp \frac{\alpha_1 F \eta}{RT} - \frac{c_{\text{int}}(\eta)}{c^0_{\text{int}}} \exp \left(- \frac{(1-\alpha_1)F\eta}{RT} \right) \right] \tag{17.10}$$

$$j_2(\eta) = j_{0,2} \left[\frac{c_{\text{int}}(\eta)}{c^0_{\text{int}}} \exp \frac{\alpha_2 F \eta}{RT} - \exp \left(- \frac{(1-\alpha_2)F\eta}{RT} \right) \right] \tag{17.11}$$

where $c^0_{\text{int}} = c_{\text{int}}(\phi_0)$, and $j_{0,1}, j_{0,2}$ denote the exchange current densities of the two reactions at the equilibrium potential. From these equations $c_{\text{int}}(\eta)/c^0_{\text{int}}$ can be eliminated so that we obtain the current-potential relation:

$$j = \frac{2 j_{0,1} j_{0,2}}{j_m} \left[\exp \frac{(\alpha_1 + \alpha_2)F\eta}{RT} - \exp \left(- \frac{(2 - \alpha_1 - \alpha_2)F\eta}{RT} \right) \right] \tag{17.12}$$

where

$$j_m = j_{0,2} \exp \frac{\alpha_2 F \eta}{RT} + j_{0,1} \exp \left(- \frac{(1-\alpha_1)F\eta}{RT} \right)$$

For high anodic or cathodic overpotentials one of the partial current densities can be neglected:

$$j = 2 j_{0,1} \exp \frac{\alpha_1 F \eta}{RT}, \quad \text{for } F\eta \gg RT \tag{17.13}$$

$$j = -2 j_{0,2} \exp \left(- \frac{(1-\alpha_2)F\eta}{RT} \right), \quad \text{for } F\eta \ll RT \tag{17.14}$$

So a Tafel plot results in straight lines at high overpotentials (see Fig. 17.1), but the two branches give different apparent exchange densities, $2j_{0,1}$ and $2j_{0,2}$, when they are extrapolated to zero overpotential. Also, the two apparent transfer coefficients obtained from the slopes do not necessarily add up to unity or to a positive integer. If the two exchange current densities differ by orders of magnitude, there is an intermediate range of potentials with a different apparent transfer coefficient, and a change in slope at high absolute values of the overpotential (see curve (b) in Fig. 17.1, and also Problem 1). Recall that we have assumed that the intermediate stays at the electrode surface. The general case where it can diffuse to the bulk of the solution is considered in Problem 2.

17.2 Electrochemical reaction order

We consider a complex reaction that contains exactly one electrochemical step of the type:

$$\mathrm{A} \rightleftharpoons \mathrm{B} + \mathrm{ze}^- \tag{17.15}$$

which is rate determining. This step can be a simple redox reaction, an ion-transfer reaction, or a metal ion deposition. We further assume that the reactants react with other species $S_1, S_2, \ldots, S_m$ through fast reactions of the form:

$$A \rightleftharpoons \sum_{i=1}^{m} x_{i,a} S_i \tag{17.16}$$

$$B \rightleftharpoons \sum_{i=1}^{m} x_{i,b} S_i \tag{17.17}$$

An example will be presented in the next section. The coefficients $x_{i,a}$ and $x_{i,b}$ are called *electrochemical reaction orders.* Usually the species A and B react only with a few of the substances S_i, so that the reaction orders for the other species vanish. We assume that the reactions (17.16) and (17.17) are in equilibrium. The total reaction is:

$$\sum_{i=1}^{m} x_{i,a} S_i \rightleftharpoons \sum_{i=1}^{m} x_{i,b} S_i + ze^- \tag{17.18}$$

Since the chemical reactions are in equilibrium, the concentrations c_a and c_b of the species A and B can be calculated from the equilibrium constants K_a and K_b and the concentrations c_i of the species S_i:

$$c_a = K_a \prod_{i=1}^{m} c_i^{x_{i,a}} \tag{17.19}$$

$$c_b = K_b \prod_{i=1}^{m} c_i^{x_{i,b}} \tag{17.20}$$

Note that we deviate slightly from the common convention according to which K_a and K_b should be the inverse of the equilibrium constants since A and B are products; our usage simplifies the notation in this context.

If the electrochemical reaction obeys the Butler–Volmer equation, the current density j at an electrode potential ϕ is:

$$\begin{aligned} j = {} & zFk^0 K_a \prod_{i=1}^{m} c_i^{x_{i,a}} \exp \frac{z\alpha F(\phi - \phi_{00})}{RT} \\ & - zFk^0 K_b \prod_{i=1}^{m} c_i^{x_{i,b}} \exp\left(-\frac{z(1-\alpha)F(\phi - \phi_{00})}{RT}\right) \end{aligned} \tag{17.21}$$

Figure 17.2 shows a set of current-potential curves, where the concentration of one of the species S_i has been varied. In the two linear Tafel regions, far from

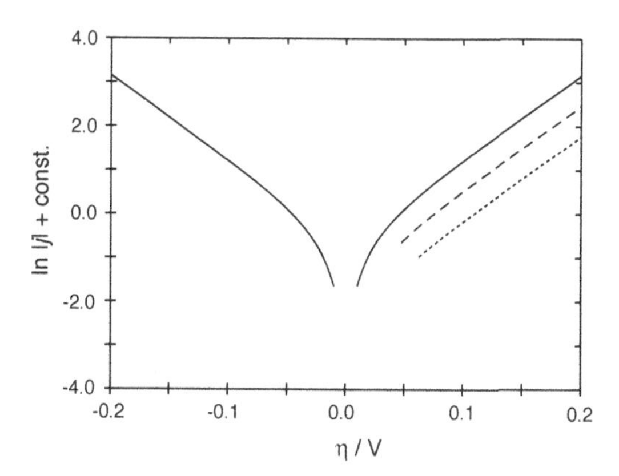

Fig. 17.2. Tafel plot for various concentrations of reactant S in equilibrium with A and electrochemical reaction order $x_a = 1$; the *dashed curve* refers to a concentration of S lowered by a factor of two, the *dotted curve* to a concentration lowered by a factor of four with respect to the *full curve*.

equilibrium, the back reactions can be neglected, so that the anodic current density is given by:

$$\ln j = \ln\left(zFk^0K_a\right) + \sum_{i=1}^{m} x_{i,a} \ln c_i + \frac{z\alpha F(\phi - \phi_{00})}{RT} \tag{17.22}$$

and the cathodic current density:

$$\ln|j| = \ln\left(zFk^0K_b\right) + \sum_{i=1}^{m} x_{i,b} \ln c_i - \frac{z(1-\alpha)F(\phi - \phi_{00})}{RT} \tag{17.23}$$

If one of the concentrations c_i is varied, the Tafel lines are shifted, and the electrochemical reaction orders $x_{a,i}$ and $x_{i,b}$ can be determined from:

$$x_{i,a} = \left(\frac{\partial \ln j}{\partial \ln c_i}\right)_{\phi, c_{i \neq j}}, \qquad \text{anodic branch} \tag{17.24}$$

$$x_{i,b} = \left(\frac{\partial \ln |j|}{\partial \ln c_i}\right)_{\phi, c_{i \neq j}}, \qquad \text{cathodic branch} \tag{17.25}$$

where all other variables, including the potential ϕ, must be kept constant.

Alternatively the electrochemical reaction orders can be determined from the exchange current density j_0. From Eq. (17.21):

$$\begin{aligned} \ln j_0 &= \ln\left(zFk^0K_a\right) + \sum_{i=1}^{m} x_{i,a} \ln c_i + \frac{z\alpha F(\phi_0 - \phi_{00})}{RT} \\ &= \ln\left(zFk^0K_b\right) + \sum_{i=1}^{m} x_{i,b} \ln c_i \\ &\quad - \frac{z(1-\alpha)F(\phi_0 - \phi_{00})}{RT} \end{aligned} \tag{17.26}$$

By differentiation we obtain:

$$\left(\frac{\partial \ln j_0}{\partial \ln c_i}\right)_{c_{i \neq j}} = x_{i,a} + \frac{z\alpha F}{RT}\,\frac{\partial \phi_0}{\partial \ln c_j} \tag{17.27}$$

$$= x_{i,b} - \frac{z(1-\alpha)F}{RT}\,\frac{\partial \phi_0}{\partial \ln c_j} \tag{17.28}$$

We need the Nernst equation to determine the change of the equilibrium potential with concentration. For this purpose the overall reaction is usually rewritten in such a way that all coefficients are integers, with negative stochiometric coefficients denoting the reactants. This results in an equation of the form:

$$0 = \sum_{i=1}^{m} \nu_i S_i + ne^- \tag{17.29}$$

where the coefficients ν_i are related to the reaction orders by:

$$\nu_i = (x_{i,b} - x_{i,a})\,\frac{n}{z} \tag{17.30}$$

The Nernst equation is then:

$$\phi_0 = \phi_{00} + \frac{RT}{nF}\sum_{i=1}^{m} \nu_i \ln c_i \tag{17.31}$$

Differentiating and substituting into Eq. (17.27) gives:

$$\frac{\partial \ln j_0}{\partial \ln c_i} = x_{i,a} + \alpha\nu_i\frac{z}{n} = x_{i,b} - (1-\alpha)\nu_i\frac{z}{n} \tag{17.32}$$

The quantities α, z, n can be determined separately, so that Eq. (17.32) offers an alternative way of obtaining the electrochemical reaction orders. A good discussion of the coupling of electrochemical with chemical reactions has been given by Parsons [1].

17.3 Mixed potentials and corrosion

The absence of a net current does not necessarily mean that the interface is in equilibrium. In fact, several reactions may proceed in such a way that the total current vanishes. We consider the case where two reactions, an anodic and a cathodic one, balance. The reaction scheme is:

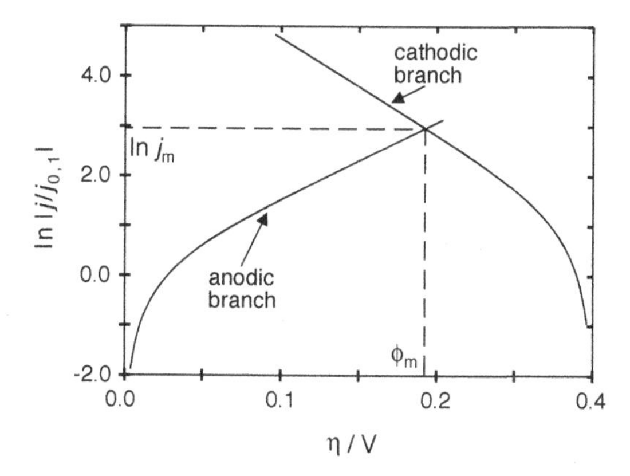

Fig. 17.3. The mixed potential originating from a cathodic and an anodic reaction.

$$\mathrm{A} \rightarrow \mathrm{B} + z_1 e^- \tag{17.33}$$

$$\mathrm{C} + z_2 \mathrm{e}^- \rightarrow \mathrm{D} \tag{17.34}$$

We assume that both reactions obey the Butler-Volmer equation, and denote the corresponding transfer coefficients by α_1 and α_2, the exchange current densities by $j_{0,1}$ and $j_{0,2}$, and the equilibrium potentials by $\phi_0^{(1)}$ and $\phi_0^{(2)}$. Since the total current density is zero we have:

$$j_{0,1} \exp \frac{z_1 \alpha_1 F \left(\phi_m - \phi_0^{(1)}\right)}{RT} = -j_{0,2} \exp\left(-\frac{z_2(1-\alpha_2)F(\phi_m - \phi_0^{(2)})}{RT}\right) \tag{17.35}$$

where ϕ_m, the potential at which there is no current, is called the *mixed potential*. We have assumed that $|\phi_1 - \phi_2| \gg RT$ so that the back reactions can be neglected. A short calculation gives for the mixed potential:

$$\phi_m = \frac{(RT/F)\ln(j_{0,2}/j_{0,1}) + z_1\alpha_1\phi_0^{(1)} + z_2(1-\alpha_2)\phi_0^{(2)}}{z_1\alpha_1 + z_2(1-\alpha_2)} \tag{17.36}$$

Each reaction proceeds with a current density of:

$$j_m = j_{0,1} \exp \frac{z_1\alpha_1 F(\phi_m - \phi_0^{(1)})}{RT} \tag{17.37}$$

Of course, one can substitute ϕ_m from Eq. (17.36), but the resulting expression is complicated. The mixed potential and the two partial currents are illustrated in Fig. 17.3.

An important example is the corrosion of metals. Most metals are thermodynamically unstable with respect to their oxides. In the presence of water or moisture, they tend to form a more stable compound, a process known as *wet corrosion* (dry corrosion is not based on electrochemical reactions and will not be considered here). Moisture is never pure water, but contains at least dissolved oxygen, sometimes also other compounds like dissolved salt. So a corroding metal can be thought of as a single electrode in contact with an aqueous solution. The fundamental corrosion reaction is the dissolution of the metal according to:

$$M \rightarrow M^{z+} + ze^- \tag{17.38}$$

This reaction can only proceed if the electrons are consumed by a cathodic counter reaction, because otherwise the metal surface would accumulate charge. Common reactions are the hydrogen evolution reaction, which in acid solutions proceeds according to:

$$2\mathrm{H}^+ + 2\mathrm{e}^- \rightarrow \mathrm{H}_2 \tag{17.39}$$

or oxygen reduction:

$$\mathrm{O}_2 + 4\mathrm{e}^- + 4\mathrm{H}^+ \rightarrow 2\mathrm{H}_2\mathrm{O} \tag{17.40}$$

For the corresponding equations in alkaline solutions, see Chap. 13 and 14. The metal surface attains a mixed potential ϕ_{cor}, the *corrosion potential*, such that the anodic current of the metal dissolution is exactly balanced by the cathodic current of one or more reduction reactions. The corrosion potential is given by Eq. (17.36), and the *corrosion current density* by Eq. (17.37).

On an inhomogeneous surface the two currents densities may vary over the surface, and need not balance locally; only the total current must be zero. In this case we must replace the exchange current densities in Eqs. (17.35), (17.36), and (17.37) by the corresponding exchange currents. Because of charge conservation an uneven current distribution on the electrode must be balanced by currents flowing parallel to the surface on both sides of the interface.

Problems

1. Consider the reaction with two consecutive electron-transfer steps described by Eq. (17.12). (a) Show that, if $j_{0,2} \gg j_{0,1}$, there is an intermediate range of negative overpotentials in which the apparent transfer coefficient is $(2 - \alpha_1)$ and the apparent exchange current density $2j_{0,1}$ (see Fig. 17.1). (b) Derive the form of the Tafel plot for $j_{0,1} \gg j_{0,2}$.
2. Consider the reaction scheme of Eq. (17.11) and assume that the intermediate can diffuse away from the electrode surface. In the simplest case the current density of particles diffusing away is proportional to the concentration of the intermediate c_{int} at the surface: $j_{\mathrm{diff}} = kc_{\mathrm{int}}$. Derive an expression for c_{int} under stationary conditions.
3. Derive Eq. (17.35) from Eqs. (17.33) and (17.34).

References

1. R. Parsons, *Trans. Faraday Soc.* **47** (1951) 1332.

18

Liquid–liquid interfaces

18.1 The interface between two immiscible solutions

When we defined electrochemistry in Chap. 1, we made a special case for including the interface between two liquid electrolytes. Because they show many similarities with the more usual electrochemical systems. Much of the interest in these interfaces resides in the fact that they can serve as models for membranes, but they are also interesting systems in their own right. Often, these interfaces are also denoted as *interfaces between two immiscible solutions* (ITIES). However, since in most cases the two liquids are not totally immiscible, we shall not use this terminology.

Most of the liquid–liquid interfaces that have been studied involve water and an organic solvent such as nitrobenzene or 1,2-dichloroethane (1,2-DCE). Although these systems form stable interfaces, the solubility of one solvent in the other is usually quite high. For example, the solubility of water in 1,2-DCE is 0.11 M, and that of 1,2-DCE in water is 0.09 M. So each of the two liquid components is a fairly concentrated solution of one solvent in the other. It is therefore unlikely that the interface is sharp on a molecular level. We rather expect an extended region with a thickness of the order of a few solvent diameters, over which the concentrations of the two solvents change rapidly (see Fig. 18.1). The lower the solubility of one solvent in the other, the thinner this interfacial region should be. However, the interfacial region does not consist of a random mixture of the two kind of solvent molecules, but rather exhibits a fluctuating boundary between the two liquids, as shown in Fig. 18.2. When this is averaged in the direction parallel to the interface, overlapping distributions of the two solvent molecules result as they are shown in Fig. 18.1. The higher the miscibility of the two solvents, the lower is the interfacial tension, and the larger are the fluctuations. In recent years, the width of the interfacial region between several immiscible liquids has been measured [1] by X-ray reflectivity, and typical values are of the order of 6 – 12 Å, in line with our expectations.

W. Schmickler, E. Santos, *Interfacial Electrochemistry*, 2nd ed.,
DOI 10.1007/978-3-642-04937-8_18, © Springer-Verlag Berlin Heidelberg 2010

The undulating structure of the interface has been described as *capillary waves*. These are waves at liquid surfaces or interfaces, where the restoring force is not gravity like in ordinary water waves, but the surface or interfacial tension. The creation of a wave at a liquid–liquid interface increases the area of the interface, and hence its energy. For a given amplitude, short waves create a larger surface area than long waves, and are therefore less likely to occur. These ideas can be put into a quantitative theory [2] and applied to liquid–liquid interfaces. However, some care must be exercised. Capillary wave theory neglects molecular structure, assumes that the amplitude of the waves are much smaller than their wavelengths, and thus holds at macroscopic and mesoscopic scales [3]. Therefore, the straightforward application of capillary wave theory to molecular structures is problematic and cannot provide quantitative results.

Many of the processes that are familiar from ordinary electrochemistry have an analog at liquid–liquid interfaces; so these form a wide field of study. We limit ourselves to a brief introduction into a few important topics: thermodynamics, double-layer properties, charge-transfer reactions, and at the end of this chapter we present a simple model for liquid–liquid interfaces. Further details can be found in several good review articles [4–7].

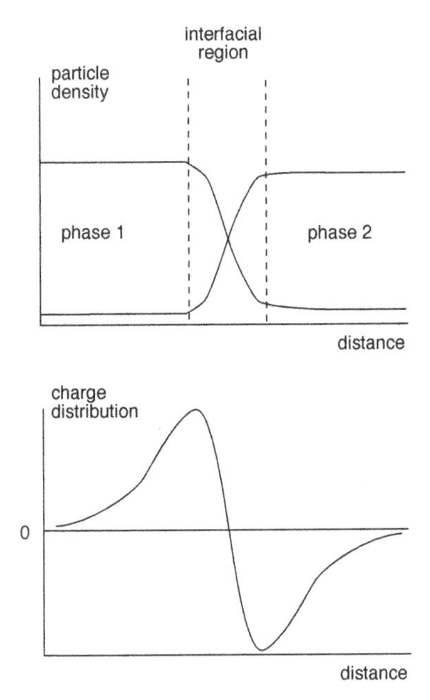

Fig. 18.1. Distribution of particles and charge at the interface between two electrolytes.

18.2 Partitioning of ions

When a solute is added to one of the two solutions, it will diffuse into the other, and after a certain time equilibrium will be established. The distribution of the solute between the two phases is known as *partitioning*, and can be determined from standard thermodynamics.

When the two solutions are in equilibrium the electrochemical potential of each species must be the same in both phases:

$$\tilde{\mu}_1 = \tilde{\mu}_2 \quad \text{or} \quad \mu_1^0 + kT \ln a_1 + ze_0\phi_1 = \mu_2^0 + kT \ln a_2 + ze_0\phi_2 \tag{18.1}$$

where μ^0 denotes the standard chemical potential, a the activity, z the charge number of the species, and the indices refer to the two adjoining solutions. For an uncharged species this results in the simple relation:

$$\frac{a_1}{a_2} = \exp \frac{\mu_1^0 - \mu_2^0}{kT} \tag{18.2}$$

The difference in the standard chemical potentials is also known as the *standard Gibbs energy of transfer*, $\Delta G_t^0 = \mu_2^0 - \mu_1^0$, since it is the Gibbs energy gained when a single particle is transferred from one solution to the other when both are in the standard state. It is determined by the difference in the energies of solvation. Note that each solvent is saturated with the other; so the standard states refer to the situation where solvent 1 is saturated with solvent 2, and vice versa. To distinguish this from the situation where each solvent is pure, it is more precise to speak of the *standard Gibbs energy of partition.*

The partitioning of ions is not so simple, since each solution must be electrically neutral with the exception of a thin boundary layer at the interface.

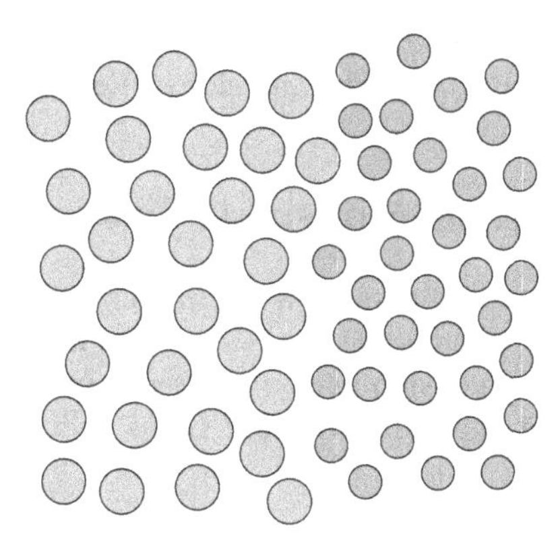

Fig. 18.2. Distribution of particles at the interface between two immiscible solvents (schematic).

As an example we consider the case where a single salt is partitioned between the two phases; for simplicity we assume that the cation and the anion have the same charge number z. We denote the cation by the index $+$, and the anion by $-$. Applying the equilibrium condition Eq. (18.1) to both ions gives for the difference in inner potentials:

$$\begin{aligned} ze_0(\phi_2 - \phi_1) &= kT \ln \frac{a_1(+)}{a_2(+)} + \mu_1^0(+) - \mu_2^0(+) \\ &= -kT \ln \frac{a_1(-)}{a_2(-)} + \mu_2^0(-) - \mu_1^0(-) \end{aligned} \tag{18.3}$$

From this we obtain for the activities:

$$\frac{a_1(+)a_1(-)}{a_2(+)a_2(-)} = \exp \frac{\mu_2^0(+) - \mu_1^0(+) + \mu_2^0(-) - \mu_1^0(-)}{kT} \tag{18.4}$$

Since each solution must be electrically neutral anions and cations have the same concentrations in the bulk:

$$a_i(+) = c_i f_i^+ \quad a_i(-) = c_i f_i^- \qquad \text{for } i = 1, 2 \tag{18.5}$$

This gives for the partitioning of the salt:

$$\frac{c_1}{c_2} = \frac{f_2^\pm}{f_1^\pm} \exp \frac{\mu_2^0(+) - \mu_1^0(+) + \mu_2^0(-) - \mu_1^0(-)}{2kT} \tag{18.6}$$

where $f^\pm = (f^+ f^-)^{1/2}$ denotes the *mean ionic activity coefficient* of the salt.

All quantities in Eq. (18.6) are measurable: The concentrations can be determined by titration, and the combination of chemical potentials in the exponent is the standard Gibbs energy of transfer of the salt, which is measurable, just like the mean ionic activity coefficients, because they refer to an uncharged species. In contrast, the difference in the inner potential is not measurable, and neither are the individual ionic chemical potentials and activity coefficients that appear on the right-hand side of Eq. (18.3).

18.3 Energies of transfer of single ions

Although the inner potential difference is not measurable in principle, it would be useful to have at least good estimates. We can see from Eq. (18.3) that this problem is equivalent to determining the difference in the chemical potential of individual ions. If we knew the *standard Gibbs energies of transfer* of the ions:

$$\Delta G_t^0(+) = \mu_2^0(+) - \mu_1^0(+); \qquad \Delta G_t^0(-) = \mu_2^0(-) - \mu_1^0(-) \tag{18.7}$$

we could calculate the inner potential difference at least in the limit of infinite dilution, where the activity coefficients are unity. For higher concentrations

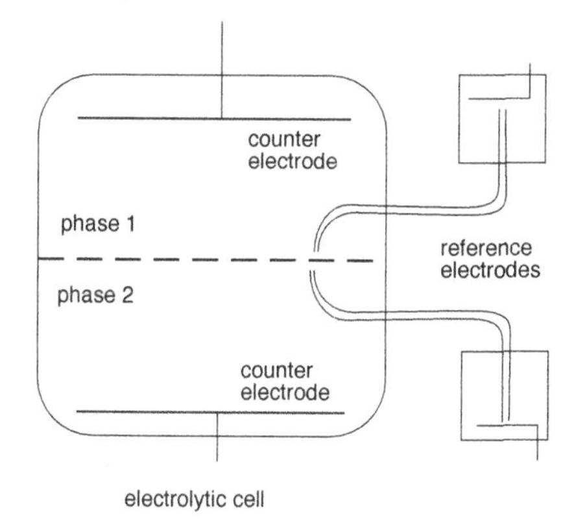

Fig. 18.3. Four-electrode configuration for liquid–liquid interfaces.

one needs an additional assumption about the activity coefficients of the ions [5]. For example, one can estimate them from the extended Debye-Hückel theory or similar models.

The Gibbs energy of transfer of a salt is measurable. If we can divide this into individual ionic contributions for one particular salt, the problem is solved for all salts, as can be seen from the following simple consideration. Suppose we had successfully divided the energy of transfer of the salt MA into the contributions of the ions M^+ and A^-. The standard Gibbs energies of transfer of some other ions N^+ and B^- are then obtained from the energies of transfer of the salts MB and NA, since these energies are additive at low concentrations, and so on for other ions. A widely used scale is based on the assumption that the energies of solvation of the tetraphenylarsonium ($TPAs^+$) and the tetraphenylborate (TPB^-) ions are equal in every solvent. This is reasonable because both ions are symmetrical, fairly large, and the charges are at the center, buried under the phenyl groups. They have, however, slightly different sizes. The resulting difference in the Gibbs energies of transfer could be estimated from the Born equation for solvation energies, but this correction is rarely made in practice. Lists of recommended values for the standard Gibbs energies of transfer can be found in the literature [4].

There are other ways of estimating inner potential differences. Girault and Schiffrin [8] assume that the difference in the inner potential is negligible at the pzc, because the interface consists of an extended layer, so that any dipole potentials will be small. The resulting scale of Gibbs energies of transfer agrees reasonably well with the $TPAs^+/TPB^-$ scale, if the small difference in the radii of these ions is accounted for.

In a real experiment one uses at least four electrodes (see Fig. 18.3), one counter and one reference electrode on each side, and measures the difference in potential between the two reference electrodes. In principle each reference electrode could be referred to the vacuum scale using the same procedure that was outlined in Chap. 4. However, in practice the required data are not available with sufficient accuracy. Of course, the voltage between the two reference electrodes characterizes the potential difference between the two phases uniquely. It can be converted to an (estimated) scale of inner potential differences by using the energies of transfer of the ions involved.

18.4 Double-layer properties

When we discussed the double-layer properties of metal electrodes in contact with an electrolyte solution, we introduced the notion of an *ideally polarizable interface*, which is marked by the absence of charge-transfer reactions over a certain *potential window* (see Chap. 5). A similar situation can prevail at liquid–liquid interfaces. Consider the interface between water and an organic solvent. If we add a strongly hydrophobic salt to the organic solvent, and for the aqueous phase use a salt that is practically insoluble in the organic phase, then there exists a potential window in which the ion transfer through the interface is negligible. Of course, in theory each salt will have a finite concentration in each solvent. However, in practice this can be entirely negligible, just as the dissolution of gold into water is negligible over a certain range of potentials.

It is natural to extend the Gouy–Chapman theory to ideally polarizable liquid–liquid interfaces. In general excess charge densities σ and $-\sigma$ exist on the two sides of the interface (see Fig. 18.1). The mathematical treatment follows the same line as for metal electrodes, but we now have two space-charge regions, one on each side of the interface. We focus on the interfacial capacity, a quantity that is accessible to experiment. The capacity C per unit area of the interface is given by the change in the charge density σ with the change in the inner potential:

$$C = \frac{d\sigma}{d(\phi_2^\infty - \phi_1^\infty)} \tag{18.8}$$

where we have added superscripts ∞ to the inner potentials to indicate that these are the limiting values far from the interface. The change in potential that is actually measured is the difference in the potential of the two reference electrodes, but this differs from $\phi_2^\infty - \phi_1^\infty$ by a constant, which drops out on differentiation. The arrangement of charges can be considered as two capacitors in series; so we may write:

$$\frac{1}{C} = \frac{1}{C_1} + \frac{1}{C_2} \tag{18.9}$$

The capacitances C_1 and C_2 of the two phases can be obtained from the Gouy–Chapman theory treated in Chap. 5. We only have to note that the potentials in the bulk of the two phases are not zero (we could set one of them equal to zero). So we replace $\phi(0)$ in Eq. (5.11) by $\phi_i^s - \phi_i^\infty$, where $i = 1, 2$, and ϕ_i^s denotes the potential in phase i at the interface. This gives for a z–z electrolyte:

$$C_i = \epsilon_i \epsilon_0 \kappa_i \cosh \frac{z_i e_0 (\phi_i^s - \phi_i^\infty)}{2kT} \tag{18.10}$$

The inner potentials ϕ_i^s have to be calculated by solving the Poisson-Boltzmann equations for the potentials in the same way as in Chap. 5.

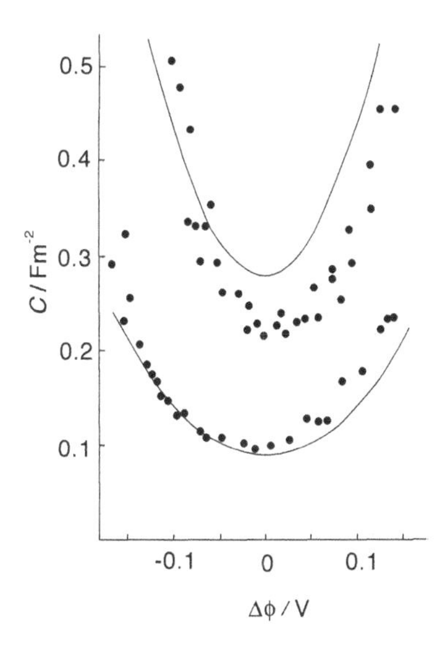

Fig. 18.4. Capacity of the interface between a solution of NaBr in water and TBAsTPB in nitrobenzene. The *upper points* are for 0.1 M solutions, the *lower* for 10^{-2} M in both phases. The *two curves* have been calculated from the Gouy-Chapman theory. The sign convention for the potential is: $\Delta\phi = \phi_w - \phi_o + \text{const.}$, where the index w stands for the aqueous and o for the organic phase. Data taken from [4].

The potentials ϕ_i^s on the two sides of the interface can differ by an interfacial dipole potential. If this changes with the applied potential it gives an extra contribution to the interfacial capacity, and Eq. (18.9) must be replaced by:

$$\frac{1}{C} = \frac{1}{C_1} + \frac{1}{C_2} + \frac{d(\phi_2^s - \phi_1^s)}{d\sigma} \tag{18.11}$$

On the whole, in the absence of specific adsorption at the interface the Gouy–Chapman theory seems to work well for liquid–liquid interfaces. Figure 18.4 shows some typical capacity curves at intermediate electrolyte concentration. For the 0.01 M solutions in both phases, the agreement between experiment and theory is good, and even for 0.1 M solutions it is quite reasonable.

As expected, for the higher concentration the theory tends to overestimate the concentration, because it does not account for the finite size of the ions and of the solvent molecules.

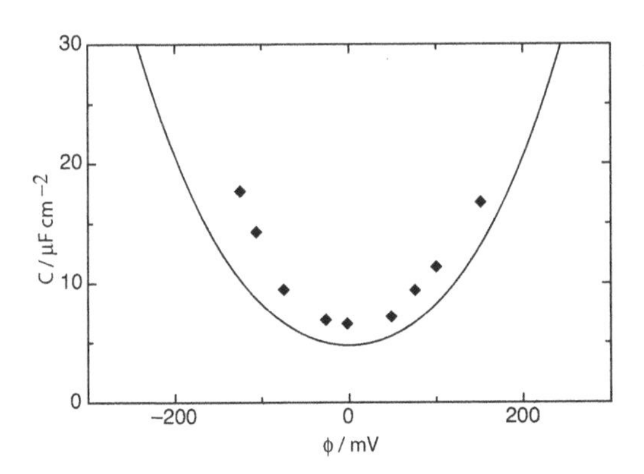

Fig. 18.5. Capacity of the interface between a 10^{-2} solution of NaBr in water and tetrabutylammonium tetraphenylborate in 1,3 dibromopropane. The *points* give the experimental values, the *curve* has been calculated from Gouy–Chapman theory. Data taken from [9].

An interesting effect is observed at low concentrations. One would expect Gouy–Chapman theory to hold even better than for higher concentrations; instead, the capacity is typically higher than predicted. A systematic study by Pereira et al. [9] on the capacity at low concentrations showed that out of ten investigated interfaces, nine had a capacity *higher* or equal to that predicted by the Gouy–Chapman theory. In contrast, on metal electrodes, the capacity, in the absence of specific adsorption, is always *lower*. An example is shown in Fig. 18.5; the experimental data are quite symmetric with respect to the pzc, which practically rules out specific adsorption (see the next paragraph). This enhancement of the capacity can be explained in terms of the diffuse structure of the interface depicted in Figs. 18.1 and 18.2. This entails on overlap of the two space-charge regions on both sides, which decreases the average separation between the opposing charges and thereby increases the capacity. Further details and model calculations can be found in [9].

Large systematic deviations from the Gouy–Chapman theory can be caused by the specific adsorption of ions at the interface. The most common cause is the pairing of ions across the interface, with one ion being in the aqueous, the other in the organic phase. As an example we mention the work of Cheng et al.[10], who studied the interface between aqueous solutions containing alkali halides and a solution of TPAsTPB in 1,2-dichloroethane. Figure 18.6 shows the capacity curves for five different alkali ions. The curves coincide at low potentials but differ significantly at higher potentials. The differences can be attributed to the pairing of the alkali cations with TPB^- at the interface. Ion pairing leads to a smaller average charge separation at the interface, and hence to a greater capacity. This effect is weakest for the Li^+ ion, and increases down the column of the periodic table: $Li^+ < Na^+ < K^+ < Rb^+ < Cs^+$. So, as may be expected, the tendency to form ion pairs at the interface is the stronger, the smaller the energy of hydration of the cation.

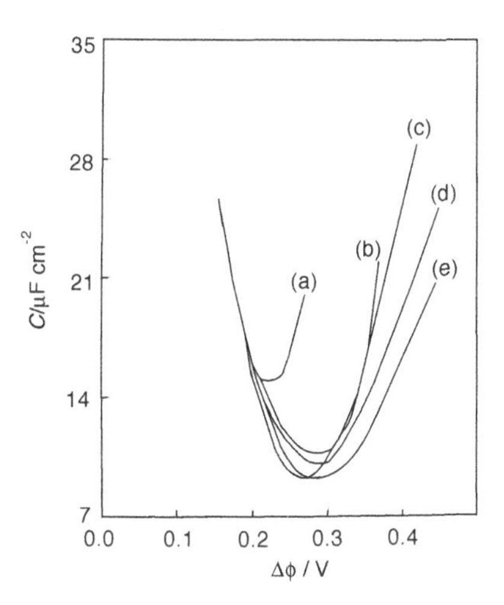

Fig. 18.6. Capacity of the interface between aqueous solutions containing alkali halides and a solution of TPAs/TPB in 1,2-dichloroethane. The electrolyte concentration in both cells was 10^{-2} M. Alkali halides used: (*a*) CsCl, (*b*) RbCl, (*c*) KCl, (*d*) NaCl, (*e*) LiCl. Data taken from [10].

18.5 Electron-transfer reactions

Electron-transfer reactions at liquid–liquid interfaces involve redox couples on each side of the interface. The basic scheme is (see Fig. 18.7):

$$\mathrm{Ox}_1 + \mathrm{Red}_2 \rightleftharpoons \mathrm{Red}_1 + \mathrm{Ox}_2 \tag{18.12}$$

where Ox_1, Red_1 are in phase 1, Ox_2, Red_2 in phase 2. Following the ideas of Sect. 4.3 we derive the equilibrium potential that is measured in the four-electrode configuration. Let phase 1 be connected to a reference electrode I, and phase 2 to reference electrode II. For simplicity, we suppose that both reference electrodes use the same metal M as electrode material. The potential drop between the two reference electrodes is:

$$\Delta\phi = \phi_{\mathrm{II}} - \phi_{\mathrm{I}} = (\phi_{\mathrm{II}} - \phi_2) + (\phi_2 - \phi_1) + (\phi_1 - \phi_{\mathrm{I}}) \tag{18.13}$$

The two reference electrodes and the interface between the two solution are in electronic equilibrium, so that we can express the differences in the inner potential through the differences in the chemical potentials. We denote the chemical potential of the two metal electrodes as μ_M, those of the two reference systems as $\mu_{\text{ref}}^{(1)}$ and $\mu_{\text{ref}}^{(2)}$, and those of the two redox couples as $\mu_{\text{redox}}^{(1)}$ and $\mu_{\text{redox}}^{(2)}$. We obtain:

$$\begin{aligned}\Delta\phi &= (\mu_M - \mu_{\text{ref}}^{(2)}) + (\mu_{\text{redox}}^{(2)} - \mu_{\text{redox}}^{(1)}) + (\mu_{\text{ref}}^{(1)} - \mu_M) \\ &= \mu_{\text{ref}}^{(1)} - \mu_{\text{ref}}^{(2)} + \mu_{\text{redox}}^{(2)} - \mu_{\text{redox}}^{(1)}\end{aligned} \tag{18.14}$$

Since systems that are in the same phase experience the same inner potential, we can write this as:

$$\Delta\phi = (\tilde{\mu}_{\text{ref}}^{(1)} - \tilde{\mu}_{\text{redox}}^{(1)}) - (\tilde{\mu}_{\text{ref}}^{(2)} - \tilde{\mu}_{\text{redox}}^{(2)}) \tag{18.15}$$

Comparison with Eq. (4.10) shows that the measured potential is simply the difference between the equilibrium potentials of the two redox couples, each measured with respect to its own reference electrode. Admittedly, this is an obvious result, but it is useful to derive it from first principles. The corresponding Nernst equation is:

$$\phi_2 - \phi_1 = \Delta\phi = \Delta\phi^0 + \frac{RT}{nF} \ln \frac{a_{\text{Red}}^{(1)} a_{\text{Ox}}^{(2)}}{a_{\text{Ox}}^{(1)} a_{\text{Red}}^{(2)}} \tag{18.16}$$

where $\Delta\phi^0$ is the standard value, when all activities are unity.

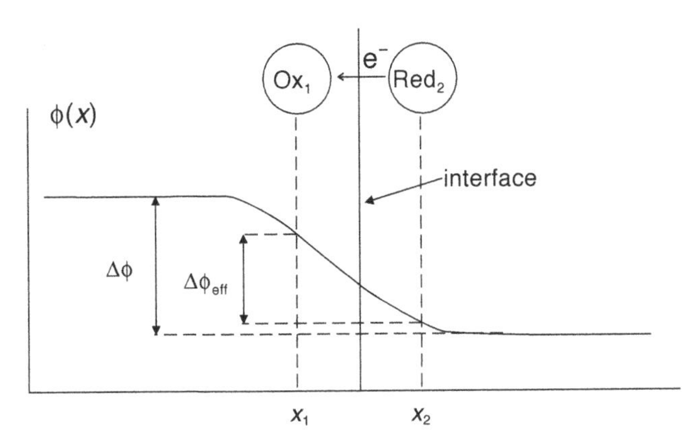

Fig. 18.7. Electron transfer at liquid–liquid interfaces. $\Delta\phi$ is the total drop in the inner potential, $\Delta\phi_{\text{eff}}$ is the part that is effective in the reaction.

Electron-transfer reactions at liquid–liquid interfaces resemble electron-transfer reactions across biological membranes, which adds a special interest.

Also, in contrast to homogeneous electron-transfer reactions, they allow a separation of the reaction products. So it is disappointing to report that only very few convincing experimental investigations of electron-transfer reactions at liquid–liquid interfaces have been performed. This is mainly due to the fact that it is difficult to find systems where the reactants do not cross the interface after the reaction; in addition, side reactions with the supporting electrolyte can be a problem.

One of the few studies that have been performed is the work of Cheng and Schiffrin [11] at the interface between water and 1,2-dichloroethane. The reactant in the aqueous phase was the $[Fe(CN)_6]^{3-/4-}$ couple, and a few different couples (e.g., lutetium diphthalocyanine) were employed in the organic phase. While the reaction rates could be measured by impedance spectroscopy (see Chap. 19), and were clearly dependent on the applied potential, an interpretation of the results is difficult. The main problem is the following: If the rate-determining step is the exchange of an electron across the interface, we need to know the variation of the electrostatic potential across the interface in order to analyze the data with the concepts familiar from electron-transfer reactions at metals. Using the notation of Fig. 18.7, from the potentials $\phi(x_1)$ and $\phi(x_2)$ at the reaction sites, we can calculate the concentrations of the reactants at the interface. The potential drop that affects the reaction rate is $\Delta\phi_{\text{eff}} = \phi(x_2) - \phi(x_1)$. Judging from the capacity data discussed above this is only a small fraction of the total potential drop $\Delta\phi = \phi_2 - \phi_1$. If we want to investigate the dependence of the reaction rate on the effective potential, we need to know how $\Delta\phi_{\text{eff}}$ varies with $\Delta\phi$. However, our double-layer theories for liquid–liquid interfaces are simply not accurate enough to furnish reliable estimates. While it would be surprising if the principles of electron-transfer reaction presented in Chaps. 8 and 9 did not hold for liquid–liquid interfaces, it is difficult to verify this. The best that can be said at the present time is that the data do not contradict the established theories. Basically, the situation has not improved much since the first edition of this book.

Recall that the situation at the interface between a metal and an electrolyte solution is much more favorable: By using a large concentration of supporting electrolyte, we can ensure that the potential at the reaction site differs little from the potential in the bulk of the solution. This does not help at liquid–liquid interfaces because for high ionic concentrations the extension of the diffuse layer is of the same order of magnitude as that of the interface itself.

In a number of cases liquid–liquid interfaces can be used to separate the products of a photoinduced electron-transfer reaction. An early example is the work by Willner et al. [12] at the water/toluene interface, who studied the photooxidation of $[Ru(bpy)_3]^{2+}$ in the aqueous phase. The excited state was quenched by hexadecyl- 4, 4′ bipyridinium, which becomes hydrophobic on reduction and crosses to the toluene phase. There are other examples and mechanisms; at the present time their main interest resides in their chemistry, and in the separation of products that can be achieved at the interface.

18.6 Ion-transfer reactions

Ion transfer across liquid–liquid interfaces is easier to study than electron transfer; so there is a greater body of experimental data. However, their interpretation is just as difficult. At the present time we can safely state:

1. Ion transfer across liquid–liquid interfaces is fast.
2. As a consequence it is difficult to separate ion transport to the interface from ion transfer across the interface.
3. There are indications that in a number of systems a Butler–Volmer-type law holds in the phenomenological sense; that is, the partial current seems to depend exponentially on the potential difference between the two phases.

As an example, Fig. 18.8 shows Tafel plots for the exchange of the acetylcholine ion between an aqueous solution and 1,2-DCE. The two branches were obtained under conditions in which the ion was initially present in one phase only. This reaction obeys the Butler–Volmer law surprisingly well, even though a microscopic interpretation faces the same difficulty that we have discussed for electron-transfer reactions.

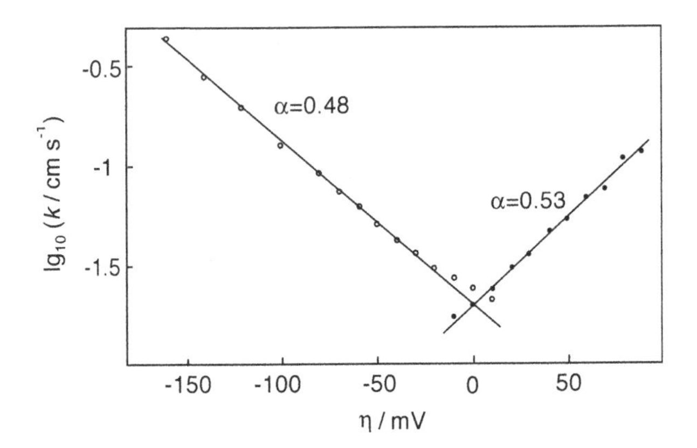

Fig. 18.8. Tafel plots for the exchange of the acetylcholine ion between an aqueous solution and 1,2-DCE; the branch on the right-hand side corresponds to transfer from the aqueous to the organic solution. Data taken from [6].

From a chemical point of view the phenomenon of *facilitated ion transfer* is intriguing. In this case, the transfer of an ion is aided by complexation in one of the phases, which shifts the equilibrium into the direction desired. Several possible mechanisms are illustrated in Fig. 18.9; for transfer from the aqueous to the organic phase, they are [6]:

ACT, aqueous complexation followed by transfer;

TOC, transfer followed by complexation in the organic phase;
TIC, transfer by interfacial complexation;
TID, transfer by interfacial dissociation.

These mechanisms are difficult to distinguish in practice, since the interface is not sharp. A good example is the facilitated transfer of sodium ions from water into 1,2-DCE [13]. The solubility of Na^+ in the organic solvent is very low, and the transfer usually requires the application of a fairly large positive potential of the aqueous with respect to the organic phase. Adding a small amount of dibenzo-18-crown-6, which acts as a *ionophore* (i.e., a complexing agent for the ion), facilitates the transfer, which then occurs at much lower potentials. The sodium ion forms a complex with the ionophore at the interface, which is then transferred to the bulk of the organic phase. By the terminology defined above this is an example of transfer by interfacial complexation.

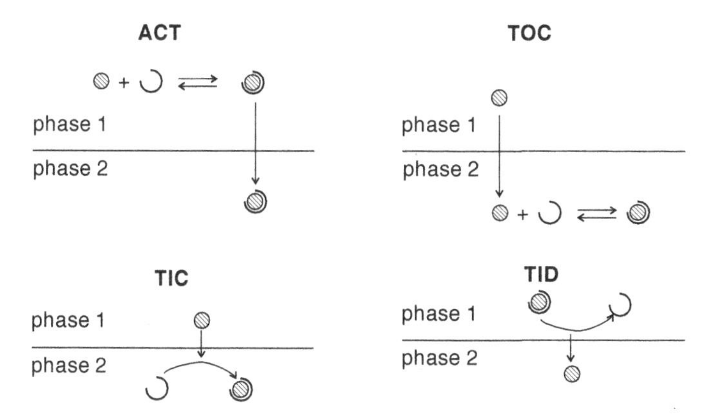

Fig. 18.9. Various mechanisms for facilitated ion-transfer reactions.

18.7 A model for liquid–liquid interfaces

Phase separations and boundaries may occur in many systems: in alloys, in ferromagnetic substances, in solutions. The basic mechanism can be understood within a simple model, which in physical chemistry is known as the lattice gas model, in magnetism as the Ising model. We present a simple version adapted to liquid–liquid interfaces.

We consider a solution composed of two types of molecules, labelled 1 and 2, which are distributed on a cubic lattice such that each lattice point is occupied by one kind of molecule. The interactions between the molecules are restricted to nearest neighbors. We denote by w_{11}, w_{22} and w_{12} the interaction energies between neighboring pairs 11, 22, and 12; they are negative for attractive interactions. The energy E of the mixture is then:

$$E = N_{11}w_{11} + N_{22}w_{22} + N_{12}w_{12} \tag{18.17}$$

where N_{ij} denotes the number of pairs ij. Let m denote the number of nearest neighbors; for a cubic lattice $m = 6$. Then the total numbers of molecules N_1 and N_2 are related to the numbers of pairs through:

$$\begin{aligned} mN_1 &= 2N_{11} + N_{12} \\ mN_2 &= 2N_{22} + N_{12} \end{aligned} \tag{18.18}$$

It is convenient to introduce the quantities:

$$\begin{aligned} E_{11} &= mN_1w_{11}/2 \\ E_{22} &= mN_2w_{22}/2 \\ w &= w_{12} - (w_{11} + w_{22})/2 \end{aligned} \tag{18.19}$$

and rewrite the total energy in the form:

$$E = E_{11} + E_{22} + N_{12}w \tag{18.20}$$

The number of pairs N_{12} is determined by the probability to find a molecule 1 at a certain site and a molecule 2 at one of the six neighboring sites. If the molecules are mixed randomly this gives:

$$N_{12} = mNx_1x_2 \quad \text{with} \quad x_1 = N_1/N \quad x_2 = N_2/N \tag{18.21}$$

As a simple approximation we will ignore any deviations from random mixing that are caused by the interactions; this procedure is also known as the *mean field approximation.*

To obtain the Helmholtz energy of the system we require the entropy, which is: $S = k \ln W$, where W is the number of different realizations of the system. The number of ways in which N_1 molecules of type 1 and $N_2 = N - N_1$ molecules of type 2 can be distributed onto N sites is:

$$W = \frac{N!}{N_1!N_2!} \tag{18.22}$$

According to Stirling's formula $\ln N! \approx N \ln N - N$; so we obtain for the entropy:

$$S = -Nk\left[x_1 \ln x_1 + x_2 \ln x_2\right] \tag{18.23}$$

Since the lattice is fixed its volume does not change with pressure, and the Gibbs and Helmholtz energies of the system are the same. Adding the energy part from Eq. (18.20) and Eq. (18.21) and the entropy part gives:

$$G = A = E_{11} + E_{22} + mNx_1x_2w + NkT\left[x_1 \ln x_1 + x_2 \ln x_2\right] \tag{18.24}$$

The last two terms give the change ΔG^M in the Gibbs energy that occurs during mixing. Using $x_2 = 1 - x_1$ we write this as:

$$\frac{\Delta G^M}{NkT} = \alpha x_1(1 - x_1) + x_1 \ln x_1 + (1 - x_1)\ln(1 - x_1) \tag{18.25}$$

where $\alpha = mw/kT$. Figure 18.10 shows this Gibbs energy of mixing as a function of the composition x_1 for various values of α. All curves are symmetric with respect to the line $x = 1/2$. Two regimes can be distinguished: If $\alpha < 2$ the Gibbs energy of mixing has a single minimum at $x_1 = 1/2$, when both components are present in the same amount. In this case any mixture of the two components will be stable. If $\alpha > 2$ the Gibbs energy of mixing has a maximum at $x = 1/2$ and two minima placed symmetrically on each side of the maximum, i.e. at positions x_1^0 and $(1 - x_1^0)$ (see Problem 4), where we may assume $x_1^0 < 1/2$. In this case the solution will separate into two phases: a phase which is richer in molecules of type 1 and a phase which contains mostly type 2. This occurs when the self-interactions w_{11} and w_{22} are much stronger than the cross-interaction w_{12}. Two solutions of different compositions are formed, which are separated by a liquid–liquid interface, so this is the case of interest to us here.

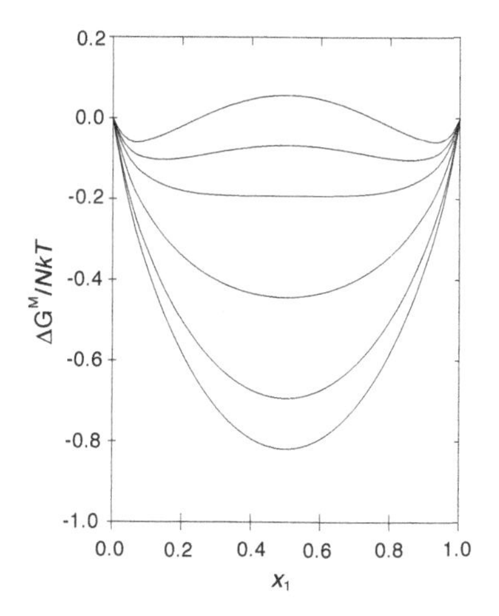

Fig. 18.10. Gibbs energy of mixing as a function of the composition; the values of α are from *top* to *bottom:* 3.0, 2.5, 2.0, 1.0, 0.0, −0.5.

At the interface the composition changes from $x_1 = x_1^0$ to $x_1 = (1 - x_1^0)$. This change is not abrupt, but occurs over an interfacial region with a certain extension in the direction perpendicular to the interface, which we would like to estimate. The thickness of this region must be determined by the condition that the Gibbs energy of forming the interface must be minimal. We first introduce the Gibbs energy $g = G^M/V$ of mixing per unit volume. In the bulk of a phase this is obtained from Eq. (18.24) by dividing through the volume and introducing the number N_V of particles per unit volume:

$$g_{\text{bulk}} = N_V \{mwx_1x_2 + kT[x_1 \ln x_1 + x_2 \ln x_2]\} \tag{18.26}$$

Near the interface the composition changes, and the Gibbs energy must contain a term depending on the rate of change. We choose our coordinate system such that the z axis is perpendicular to the interface. The leading term must involve the gradient of the composition dx_1/dz; for reasons of symmetry it has to be invariant to a change of sign, so it must be proportional to the square of the gradient. So, in the simplest approximation the Gibbs energy per volume is:

$$g(z) = g_{\text{bulk}} + \gamma \left(\frac{dx_1}{dz}\right)^2 \tag{18.27}$$

In order to estimate the coefficient γ we consider the hypothetical situation in which the composition at the interface changes abruptly from $x_1 = 1$ to $x_1 = 0$. In this case the gradient is $dx_1/dz = 1/a$, where a is the lattice constant. Compared to the situation in the bulk of the two phases a new pair 12 has been formed per surface molecule, but two bonds 11 and 22 have been broken. The excess energy per atom is obtained by the following steps: A uniform phase consisting only of molecules 1 is split in two; the change in energy per newly created surface atom is $w_{11}/2$. Similarly a phase consisting solely of molecules 2 is split, and the surface energy per atom is $w_{22}/2$. Then two half crystals with different composition are joined, and the gain in energy is w_{12} per pair of atoms. Hence the excess energy per atom at the interface is: $w_{12} - w_{11}/2 - w_{22}/2 = w$. A different derivation is given in [14].

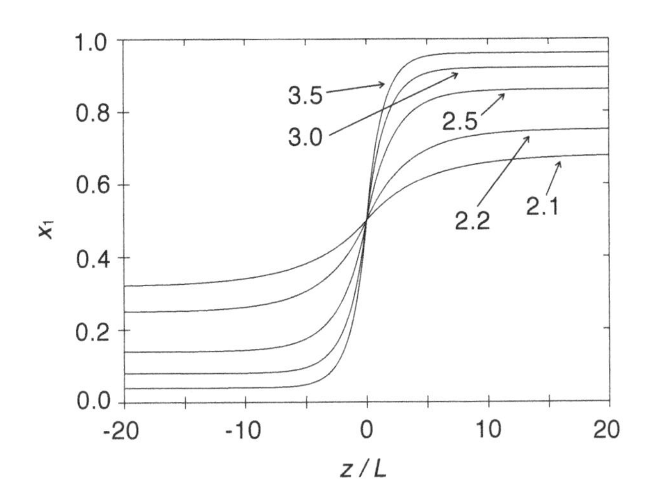

Fig. 18.11. Density profile at the interface between two immiscible solutions; the labelling indicates the values of α.

Hence:

$$\frac{g(z)}{N_V kT} = \alpha x_1 x_2 + [x_1 \ln x_1 + x_2 \ln x_2] + \alpha a^2 \left(\frac{dx_1}{dz}\right)^2 \tag{18.28}$$

The total energy of mixing

$$G^M = \int_{-\infty}^{\infty} g(z)\, dz \tag{18.29}$$

obtains its minimum for the true density profile $x_1(z)$. We derive an approximate solution by the following considerations: Far from the interface, bulk equilibrium conditions prevail. Let us take:

$$\lim_{z\to\infty} x_1(z) = x_1^0 \qquad \lim_{z\to-\infty} x_1(z) = 1 - x_1^0 \tag{18.30}$$

In a simple approximation we may assume that the density profile on either side of the interface tends exponentially towards its limiting value, and has the form:

$$x_1(z) = \begin{cases} (1-2x_0)\left[1-\frac{1}{2}\exp(z/L)\right] + x_0 & \text{for} \quad z < 0 \\ \frac{1}{2}(1-2x_0)\exp(-z/L) + x_0 & \text{for} \quad z > 0 \end{cases} \tag{18.31}$$

where the decay length L must be chosen such that the energy of mixing is minimal. This minimization can easily be achieved numerically. Within this simple model the single parameter α determines both the equilibrium composition x_0 and the decay length L/a measured in terms of the lattice constant a. Fig. 18.11 shows the density profile for different α and Table 18.1 gives a few representative values. The larger α, the smaller is the solubility of one species in the other, and the smaller the decay lenght. Conversely, the closer α gets to the critical value of $\alpha = 2$, the larger the decay length. In other words, the weaker the cross-interaction w_{12} is compared to the self-interactions w_{11} and w_{22}, the sharper is the boundary between the two phases. We have treated the lattice-gas in the mean-field approximation. Therefore the

α	x_0	L/a
2.1	0.68	4.65
2.2	0.75	3.33
2.5	0.86	2.13
3.0	0.92	1.52
3.5	0.96	1.26

Table 18.1. Bulk composition and decay length at the interface for various values of α.

density profiles that we have obtained are spatial averages. If the same model is treated more exactly, e.g. by Monte–Carlo simulations [15], the interface exhibits the same undulating structure that is schematically depicted in Fig. 18.2.

Problems

1. Consider a planar interface between two immiscible solvents with dielectric constants ϵ_1 and ϵ_2. Calculate (a) the Coulomb interaction of two ions situated on different sides of the interface; (b) the image energy of a single ion near the interface.
2. From Born's formula (cf. Eq. (3.7)) derive an expression for the difference in the energy of solvation of a spherical ion in two solvents with different dielectric constants.
3. Derive and solve the appropriate linear Poisson–Boltzmann equation for the interface between two immiscible solutions.
4. Prove that for $\alpha > 2$ the Gibbs energy of mixing has one maximum and two minima.

References

1. G. Luo et al., *Electrochem. Commun.* **7** (2005) 627, and references therein.
2. J.S. Rowlinson and B. Widom, *Molecular Theory of Capillarity.* Clarendon Press, Oxford, 1982.
3. S. Frank and W. Schmickler, *J. Electroanal. Chem.* **564** (2004) 239.
4. P. Vanysek, *Electrochemistry on Liquid–Liquid Interfaces*, Lecture Notes in Chemistry, Vol. 39. Springer, New York, NY, 1985.
5. H.H. Girault and D.J. Schiffrin, *Electroanalytical Chemistry*, Vol. 15, edited by A.J. Bard, M. Dekker. New York, NY, 1989.
6. H.H. Girault, *Modern Aspects of Electrochemistry*, Vol. 25, edited by J. O'M. Bockris et al. Plenum Press, New York, NY, 1993.
7. *Liquid Interfaces in Chemical, Biological and Pharmaceutical Applications*, edited by A.G. Volkov. Marcel Dekker, New York, NY, 2000.
8. H.H. Girault and D.J. Schiffrin, *Electrochim. Acta* **31** (1986) 1341.
9. Z. Samec, *Pure Appl. Chem.* **76** (2004) 2147.
10. C.M. Pereira, W. Schmickler, A.F. Silva, and M.J. Sousa, *Chem. Phys. Lett.* **268** (1997) 13.
11. Y. Cheng, V.C. Cunnane, D.J. Schiffrin, L. Mutomäki, and K. Kontturi, *J. Chem. Soc. Faraday Trans.* **87** (1991) 107.
12. Y. Cheng and D.J. Schiffrin, *J. Chem. Soc. Faraday Trans.* **89** (1993) 199.
13. I. Willner, W.E. Ford, J.W. Otvos, and M. Calvin, *Nature* **244** (1988) 27.
14. Y. Shao and H.H. Girault, *J. Electroanal. Chem.* **334** (1992) 203.
15. J.W. Cahn and J.E. Hilliard, *J. Chem. Phys.* **28** (1958) 258.
16. W. Schmickler, The lattice gas and other simple models for liquid–liquid interfaces. In: *Liquid Interfaces in Chemical, Biological and Pharmaceutical Applications*, edited by A.G. Volkov, Marcel Dekker, New York, NY, 2000.

19

Experimental techniques for electrode kinetics – non-stationary methods

19.1 Overview

In electrochemical kinetics, the electrode potential is the most important variable that is controlled by the experimentalist, and the current is usually measured as the response. Ideally, one would like to measure current and potential at constant, well-determined concentrations of the reactants. However, generally the concentrations of the reacting species at the interface are different from those in the bulk, since they are depleted or accumulated in the course of the reaction. So one must determine the interfacial concentrations. There are two principal ways of doing this. In the first class of methods one experimental variable, typically either the potential or the current, is kept constant or varied in a simple manner, the other observables are measured, and the surface concentrations are calculated by solving the transport equations under the conditions applied. In the simplest variant the overpotential or the current is stepped from zero to a constant value; the transient of the other variable is recorded and extrapolated back to the time at which the step was applied, when the interfacial concentrations were not yet depleted. This is the class of techniques that we cover in this chapter.

In the other class of method the transport of the reacting species is enhanced by convection. If the geometry of the system is sufficiently simple, the mass transport equations can be solved, and the surface concentrations calculated. They will be treated in the following chapter.

Besides the potential and current steps mentioned above, there are several other methods by which the system can be perturbed; the more important ones are listed in Table 19.1. Usually, at the starting point of the perturbation the system is in equilibrium. Alternatively, it can be in a stationary state, in which all the fluxes, in particular the current, are constant. If the system returns to a stationary or equilibrium state after the perturbation, one speaks of a transient technique; the first four methods in the table are of this kind. In this case, we can obtain information about different processes occurring with different velocities by analyzing the response at different time scales.

W. Schmickler, E. Santos, *Interfacial Electrochemistry*, 2nd ed.,
DOI 10.1007/978-3-642-04937-8_19, © Springer-Verlag Berlin Heidelberg 2010

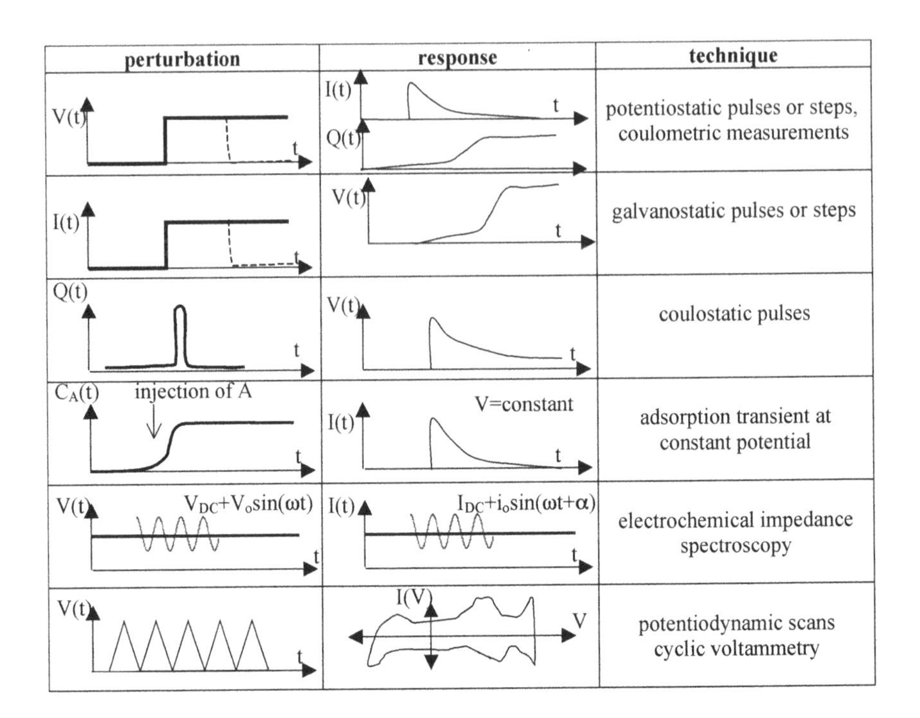

Table 19.1. Overview over the various perturbation methods; $V(t)$ denotes the potential, $I(t)$ the current, $Q(t)$ the charge, and $C_A(t)$ the concentration of a reacting species.

Alternatively, the perturbation can be periodic, and after an initial, transitory period the response will be periodic as well. This is true for the last two methods listed. In this case, the variation of the frequency of the perturbation is the key to studying processes occurring at different velocities. If the amplitude of the perturbation is small enough, the Butler–Volmer equation can be linearized and the current is proportional to the potential at the interface. Also in the case of periodically perturbation, the frequency of the response is the same as that of the perturbation. When the perturbation is large, we speak about a nonlinear response. In summary, in order to investigate the kinetics of different processes occurring at an electrochemical interface we first have to determine the potential region at which the process of interest occurs, and secondly we have to tune the time scale or frequency with the time constants of the process.

In simple cases the measured current is proportional to the rate of an electrochemical reaction. The interpretation becomes complicated if several reactions take place simultaneously. Since the measured current gives only the sum of the rate of all charge-transfer reactions, the elucidation of the reaction mechanism and the measurement of several rate constants becomes an art. A number of tricks can be used, such as complicated potential or current programs, auxiliary electrodes etc., which work for special cases.

There are several good books on the classical electrochemical techniques [1–6]. Here we give a brief outline of the most important methods. We mostly restrict ourselves to the study of simple reactions, but will consider one example in which the charge-transfer reaction is preceded by a chemical reaction.

19.2 Effect of mass transport and charge transfer on the current

Generally the current density j that is measured is determined both by the rate of the electrochemical reaction and by the transport of the reacting species to the interface. Since both processes are in series, the slower of them determines the overall current. From an electrochemist's point of view there is little interest in transport processes as such, and we would like to eliminate their effect on the data. For this purpose it is convenient to define a few quantities.

If transport were infinitely fast, the concentrations c^s_{ox} and c^s_{red} of nonadsorbing reacting species would be the same at the interface as in the bulk. The measured current density would solely be determined by the reaction, and would equal the *kinetic current density*:

$$j_k = nF(k_{\text{ox}}c^0_{\text{red}} - k_{\text{red}}c^0_{\text{ox}}) \tag{19.1}$$

where c^0 denotes a bulk concentration, and n is the number of electrons transferred ($n = 1$ for outer-sphere electron-transfer reactions). In the case of a simple reaction obeying the Butler–Volmer equation j_k is given by Eqs. (9.10) and (9.13) with $c^s = c^0$.

The other limiting case is that of an infinitely fast reaction, when the current is determined by transport only. It is customary to call such a reaction *reversible*, and denote the corresponding current density, which is determined by transport alone, as the *reversible current density* j_{rev}. It is determined by the transport, usually by diffusion, because right at the electrode surface transport of the reacting species is by diffusion alone – convection cannot carry a species right to the surface because the component of the solution flow perpendicular to the surface must vanish. One also speaks of a *diffusion current density* j_d in this case. It is obtained from the following considerations: If the reaction is infinitely fast, the electrode is in equilibrium with the reacting species at the interface; hence the concentrations c^s_{ox} and c^s_{red} are determined solely by Nernst's equation. The current is obtained by solving the diffusion equation with these surface concentrations as boundary conditions. The diffusion current density is then obtained from:

$$j_d = -z_i F D_i \left(\frac{dc_i}{dx}\right)_{x=0} \tag{19.2}$$

where x is the coordinate in the direction perpendicular to the surface, which is situated at $x = 0$, i denotes the reacting species, D_i its diffusion coefficient, and z_i its charge number. In the special case when the surface concentration of the reacting species is negligibly small compared with c^0, we speak of a *diffusion limited current density* $j_{\rm lim}$; under these conditions every molecule of species i arriving at the surface is immediately consumed.

Since transport and electrochemical reactions are in series, the slower process determines the overall current. Hence we can obtain the rate constants of the reaction only, if the reversible current $j_{\rm rev}$ is not much slower than the kinetic current. This limits the magnitude of the reaction rates that can be measured with any given method.

19.3 Potential step

The principle of this method is quite simple: The electrode is kept at the equilibrium potential at times $t < 0$; at $t = 0$ a potential step of magnitude η is applied with the aid of a potentiostat (a device that keeps the potential constant at a preset value), and the current transient is recorded. Since the surface concentrations of the reactants change as the reaction proceeds, the current varies with time, and will generally decrease. Transport to and from the electrode is by diffusion. In the case of an infinitely fast reaction (reversible reaction), i.e. when the potential is stepped to a region that is far removed from the equilibrium value, the concentration of the reacting species at the interface is reduced to zero immediately upon application of the potential step. The gradient of the concentration at the surface decreases with the inverse of the square root of time (see Fig. 19.1). Thus, the current is only determined by transport, it does not depend on the applied potential and its variation with time is given by the Cottrell equation:

$$j_{\rm rev} = \frac{zFD^{1/2}c^\circ}{\pi^{1/2}t^{1/2}} \tag{19.3}$$

In the case of a simple redox reaction obeying the Butler–Volmer law, the diffusion equation can be solved explicitly, and the transient of the current density $j(t)$ is (see Fig. 19.2, upper panel):

$$j(t) = j_k(\eta) \exp \lambda^2 t \ \text{erfc}\ \lambda t^{1/2} \equiv j_k(\eta) A(t) \tag{19.4}$$

where λ is a constant given by:

$$\lambda = \left[\frac{j_0}{F} \frac{1}{c^o_{\rm red} D^{1/2}_{\rm red}} \exp \frac{\alpha e_0 \eta}{kT} + \frac{j_0}{F} \frac{1}{c^o_{\rm ox} D^{1/2}_{\rm ox}} \exp \left(- \frac{(1-\alpha) e_0 \eta}{kT} \right) \right] \tag{19.5}$$

j_0 and α are the exchange current density and the transfer coefficient of the redox reaction, and $j_k(\eta)$ is the *kinetic current density* defined above.

At long times, for $\lambda t^{1/2} \gg 1$, we can use the asymptotic expansion of the error function, and $j \to j_{\text{rev}}$ (see Problem 2). This does not contain any information about the rate constant.

At short times, for $\lambda t^{1/2} \ll 1$, the function $A(t)$ can be expanded:

$$A(t) \approx 1 - 2\lambda\sqrt{t/\pi} \tag{19.6}$$

Under these conditions a plot of j versus $t^{1/2}$ gives a straight line, and the kinetic current can be obtained from the intercept (see Fig. 19.2, middle panel). Furthermore, the rate constant may be obtained for different overpotentials and in consequence the transfer coefficient can be also calculated. If the reaction is fast the straight portion can be too short for a reliable determination of j_k; in this case one should obtain estimates for j_k and λ from this plot, and use them in fitting the whole curve to Eq. (19.4).

A more elegant method consists in using the Laplace transform, which many mathematics packages contain as a standard option. In general, the Laplace transform $\tilde{f}(s)$ of a function $f(t)$ is defined as:

$$\tilde{f}(s) = \int_0^\infty e^{-st} f(t)\, dt \tag{19.7}$$

where the variable s has the meaning of a frequency. In the potentiostatic method, the Laplace transform of the current density takes on the simple form:

$$\tilde{j}(s) = \frac{zFc^\circ\sqrt{D}}{\sqrt{s}(k + \sqrt{sD})} \tag{19.8}$$

It is convenient to introduce an auxiliary function:

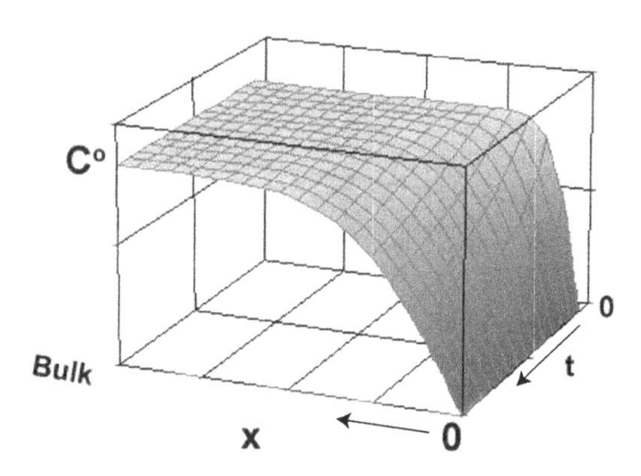

Fig. 19.1. Concentration profile for a reversible, infinitely fast, reaction. Time increases from the back towards the front, the distance from the electrode from *right* to *left*.

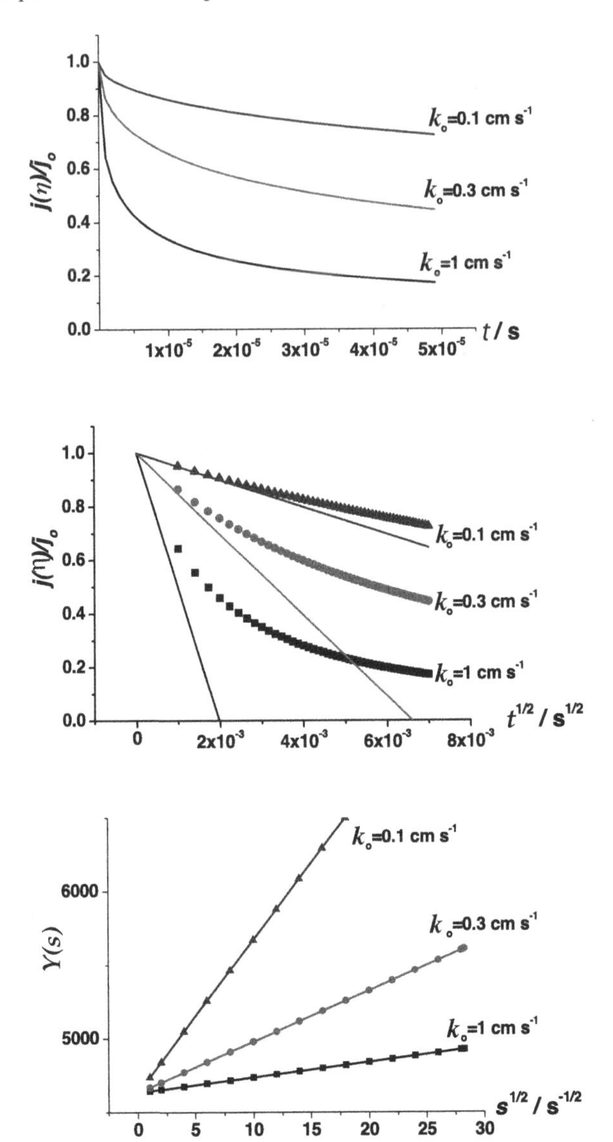

Fig. 19.2. Current transients after application of a potential step for various reaction rates (*upper panel*). Current transient plotted vs. $t^{1/2}$ (*middle panel*); the *straight lines* give the short-time limit according to Eq. (19.6). Plot of $Y(s)$ obtained by Laplace transform (*lower panel*).

$$Y(s) = \frac{1}{\tilde{j}(s)\sqrt{s}} = \frac{1}{zF\sqrt{D}c^{\circ}} + \frac{\sqrt{s}}{zFkc^{\circ}} \tag{19.9}$$

Thus, a plot of $Y(s)$ versus the square root of the frequency gives a straight line, and the rate constant can be obtained from the slope (see Fig. 19.2, lower panel). The intercept contains only parameters pertaining to mass transport. Note that the slope decreases with the rate k, and the uncertainty becomes large for fast reactions.

There are two difficulties with this method. The first one is due to the fact that in reality the potentiostat keeps the potential between the working and the reference electrode constant; there is an ohmic resistance R_Ω between the tip of the Luggin capillary (see Chap. 4) and the working electrode, giving rise to a potential drop IR_Ω (I is the current). Since I varies in time, so does the potential drop by which η is in error. However, modern potentiostats can correct for this to some extent. The second difficulty is more serious. Immediately after the potential step the double layer, which acts as a capacitor, is charged, and double layer-charging and the Faradaic current due to the reaction cannot be separated. If the reaction is fast, the surface concentrations change appreciably while the double layer is charged, and Eqs. (19.4) and (19.5) no longer hold. This limits the range of rate constants that can be determined with this method to $k_0 \leq 1$ cm s^{-1}.

19.4 Current step

A related technique is the current-step method: The current is zero for $t < 0$, and then a constant current density j is applied for a certain time, and the transient of the overpotential $\eta(t)$ is recorded (Fig. 19.3). The correction for the IR_Ω drop is trivial, since I is constant, but the charging of the double layer takes longer than in the potential step method, and is never complete because η increases continuously. The superposition of the charge-transfer reaction and double-layer charging creates rather complex boundary conditions for the diffusion equation; only for the case of a simple redox reaction and the range of small overpotentials $|\eta| \ll kT/e_0$ is the transient fairly simple:

$$\eta(t) = \frac{kT}{e_0}\left[\frac{1}{j_0} + \frac{2B}{F}\left(\frac{t}{\pi}\right)^{1/2} - RTC\left(\frac{B}{F}\right)^2\right] j \tag{19.10}$$

with:

$$B = \frac{1}{c^0_{\text{ox}} D^{1/2}_{\text{ox}}} + \frac{1}{c^0_{\text{red}} D^{1/2}_{\text{red}}} \tag{19.11}$$

where C is the double-layer capacity at the equilibrium potential. A plot of η versus $t^{1/2}$ does not give the exchange current density directly by extrapolation; the double-layer capacity must be determined separately.

These equations cannot be used at higher overpotentials $|\eta| \geq kT/e_0$. If the reaction is not too fast, a simple extrapolation by eye can be used. The potential transient then shows a steeply rising portion dominated by double-layer charging followed by a linear region where practically all the current is

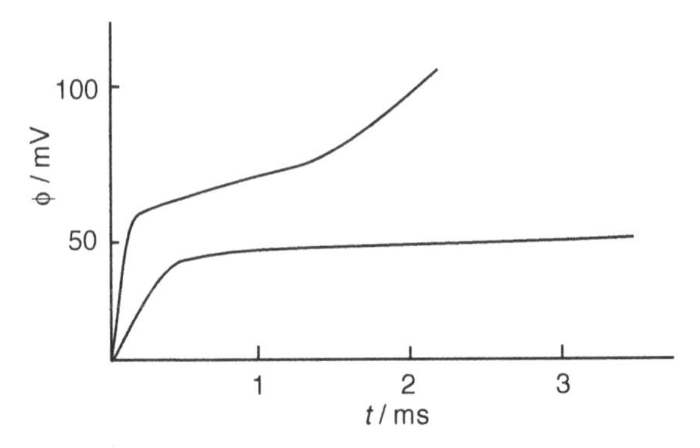

Fig. 19.3. Potential transient for a current step.

due to the reaction (see Fig. 19.3). Extrapolation of the linear part to $t = 0$ gives a good estimate for the corresponding overpotential.

If the reaction is too fast for this procedure, a *double-pulse method* can be used: The current pulse is preceded by a short but high pulse which is designed to charge the double layer. The height of the pulse is adjusted in such a way that the transient $\eta(t)$ is horizontal at the beginning of the second pulse, and this portion is then extrapolated to $t = 0$. This method is only approximate, and adjusting the height of the first pulse is tedious, but it does extend the range of application to faster reactions. Even so the current pulse method is limited to reactions with $k_0 \leq 1$ cm s^{-1} just like the potential step method.

19.5 Coulostatic pulses

In the sixties Delahay [7] and Reinmuth [8] developed the idea to measure the rates of electrochemical reactions by charging the double layer with a very short pulse. The rate constant is determined by analyzing the subsequent relaxation of the potential to the equilibrium conditions. Although this is an excellent method to measure fast reactions, it is underused.

The experimental setup is very simple and inexpensive – see Fig. 19.4. A coulostat can be home-made and consists of a condenser, a current supply and a fast relay. A condenser with a capacity C_c much smaller than that of the double layer C_d injects an amount of charge Q_0 into the cell during a very short time. If the time constant for discharging the condenser is much shorter than that for the double layer, $\tau_c = R_\Omega C_c \ll \tau_k = R_k C_d$, no leaking of charge through the charge transfer resistance R_k occurs during the pulse. Thus, at the end of the pulse the system is at open circuit and the charge accumulated at the double layer is Q_0. This is an advantage in comparison with other methods since no compensation for the electrolyte resistance is necessary.

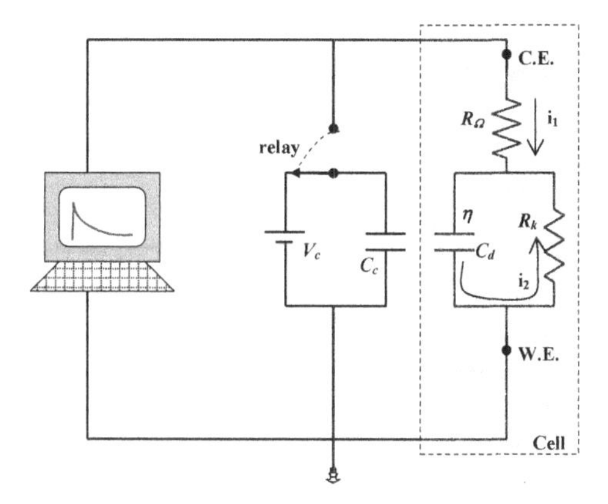

Fig. 19.4. Basic setup for the coulostatic method..

Thus at $t = 0$ the overpotential at the interface is $\eta_0 = Q_0/C_d$. When a redox couple is present (Ox/Red), the double layer subsequently discharges through the Faradaic resistance R_k and the potential decay is recorded (see Fig. 19.5, upper panel). In the absence of mass transport limitations, and if the perturbation is sufficiently small such that the Butler–Volmer equation can be linearized (see Eq. (9.15)) the transient $\eta(t)$ decays exponentially:

$$\eta(t) = \eta_0 \exp(-t/\tau_k) \tag{19.12}$$

The exchange current, and hence the rate constant, can be calculated from R_k according to:

$$j_0 = \frac{RT}{zFR_k} \tag{19.13}$$

Since the injected charge is known, the value of the double-layer capacity can be obtained from the initial overpotential η_0.

Equation (19.12) is only valid when mass transport plays no role, which is always true at very short times. Taking into account the boundary conditions and solving the Fick's laws the concentration profiles at the interface can be obtained. Figure 19.6 shows the results for different times after the application of the coulostatic pulse. We consider that a charge pulse Q_0 has been applied. The system attempts to recover the equilibrium conditions and an oxidation or reduction process starts. When the double layer begins to discharge the concentrations c_{ox} of the oxidized or c_{red} of the reduced species at the interface gradually change. However, the profile attains a maximum (or minimum) as a consequence of the depletion of the charge. This extremum shifts to longer times and becomes broader at larger distances from the interface. Then the perturbation propagates to the bulk and becomes attenuated. The more exact expression for the transient $\eta(t)$ considering the diffusion processes is now:

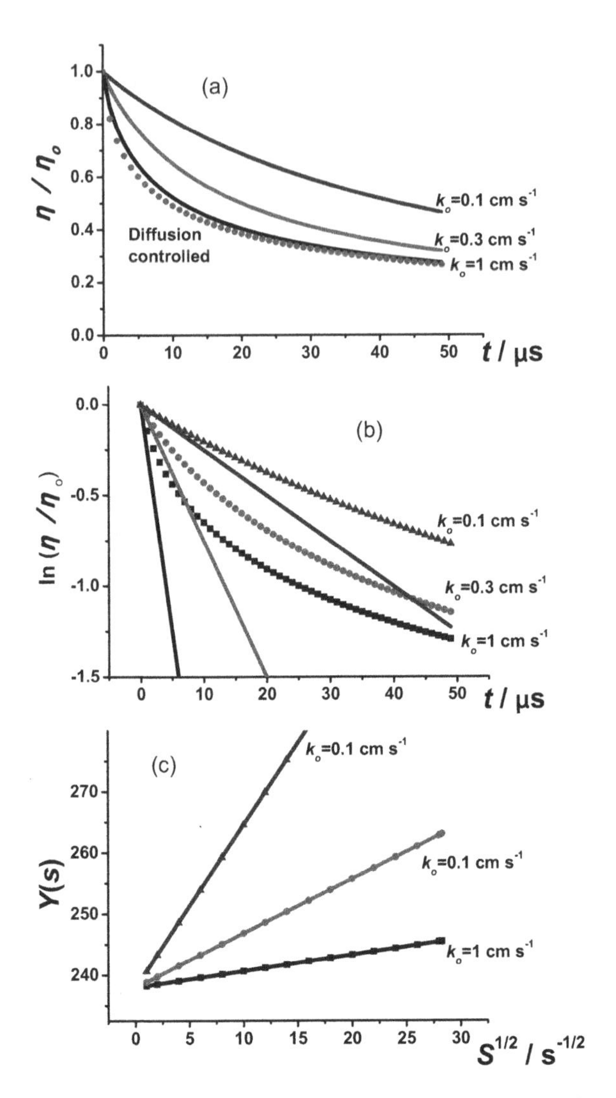

Fig. 19.5. Potential transients after application of a coulostatic pulse for various reaction rates (*upper panel*). Logarithmic plot (*middle panel*); the *straight lines* give the short-time limit according to Eq. (19.12). Plot of $Y(s)$ obtained by Laplace transform (*lower panel*).

$$\eta(t) = \eta_0 \frac{1}{b-a}\left[b\exp(a^2t)\operatorname{erfc}(at^{1/2}) - a\exp(b^2t)\operatorname{erfc}(bt^{1/2})\right] \qquad (19.14)$$

where:

$$a = \frac{\tau_d^{1/2} + (\tau_d - 4\tau_k)^{1/2}}{2\tau_k}, \qquad b = \frac{\tau_d^{1/2} - (\tau_d - 4\tau_k)^{1/2}}{2\tau_k} \qquad (19.15)$$

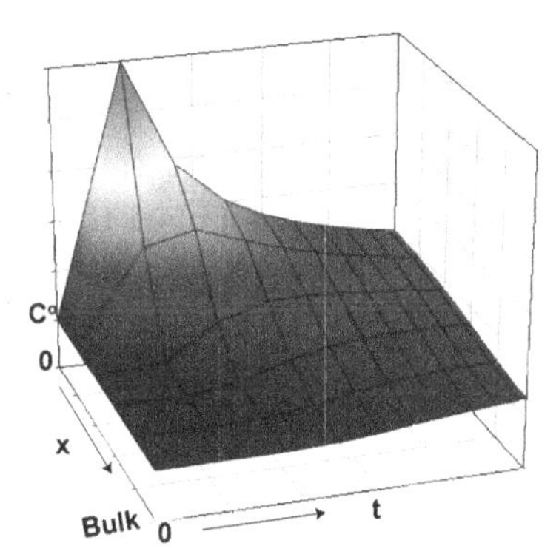

Fig. 19.6. Concentration profile after application of a coulostatic pulse.

and:

$$\tau_d^{1/2} = \frac{RTC_d}{n^2F^2}\left[\frac{1}{D_{\text{ox}}^{1/2}C_{\text{ox}}} + \frac{1}{D_{\text{red}}^{1/2}C_{\text{red}}}\right] \tag{19.16}$$

Figure 19.5a shows the normalized transients $\eta(t)/\eta_0$ for three different rate constants. For comparison, the response for an infinitly fast reaction controlled by mass transport is also shown. The limitations of the analysis using Eq. (19.12) are shown in Fig. 19.5b. A deviation from linearity is observed even for the slowest reaction at times as short as 10 μs.

Just as in the potential step method, a more convenient analysis can be performed in the frequency domain. The Laplace transform $\tilde{\eta}(s)$ of the overpotential obeys a much simpler equation. We define a function $Y(s)$ of the frequency s which correlates with the kinetic and mass transport parameters through:

$$Y(s) = \frac{1}{(Q_0/\tilde{\eta}(s) - C_d)\, s^{1/2}} = R_k s^{1/2} + \frac{\tau_d^{1/2}}{C_d} \tag{19.17}$$

The evaluation of the rate constant can be done by plotting the function $Y(s)$ against $s^{1/2}$, (see Fig. 19.5c), which results in a straight line with R_k as slope and the diffusion parameter $\tau_d^{1/2}$ divided by the double layer capacity as intercept. In this way the evaluation of the kinetic parameters is independent of the knowledge of the diffusion parameters. This would not have been the case if the data had been fitted directly with Eq. (19.14). The coulostatic method has been successfully used to determine rate constant of fast reactions and values of the same order of magnitude as those obtained by turbulent pipe flow (see next chapter) have been obtained [9].

19.6 Impedance spectroscopy

An alternative strategy to investigate electrochemical reactions is directly to work in the frequency domain. In impedance spectroscopy a sinusoidally varying potential with a small amplitude is applied to the interface, and the resulting response of the current measured. It is convenient to use a complex notation, and write the applied signal in the form:

$$V(t) = V_0 e^{\mathrm{i}\omega t} \tag{19.18}$$

where it is understood that the real part of this equation describes the physical process. When the amplitude V_0 is sufficiently small, $V_0 \ll kT/e_0$, the response of the interface is linear, and the current I takes the form:

$$I(t) = I_0 e^{\mathrm{i}\omega t} \tag{19.19}$$

where the amplitude I_0 of the current is generally complex (i.e., the current response has a phase shift denoted by $-\varphi$):

$$I_0 = \mid I_0 \mid e^{-\mathrm{i}\varphi} \tag{19.20}$$

The impedance of the system is the ratio:

$$Z = V_0/I_0 = \mid Z \mid e^{\mathrm{i}\varphi} \tag{19.21}$$

Typically, the frequency ω of the modulation is varied over a considerable range, and an *impedance spectrum* $Z(\omega)$ recorded. Various electrode processes make different contributions to the total impedance. In many cases it is useful to draw an *equivalent circuit* consisting of a number of simple elements like resistors and capacitors, arranged in parallel and in series. However, in complicated systems more than one equivalent circuit with the same overall impedance may exist, and the interpretation becomes difficult.

We consider a simple redox reaction obeying the Butler–Volmer equation. At small overpotentials, the *charge-transfer impedance* is:

$$Z_k = \frac{RT}{Fj_0} \tag{19.22}$$

Double-layer charging gives rise to an impedance:

$$Z_C = \frac{-\mathrm{i}}{\omega C_d} \tag{19.23}$$

These two impedances are in parallel. The resistance R_Ω between the working and the reference electrode is purely ohmic, and is in series with the other two.

At high frequencies diffusion of the reactants to and from the electrode is not so important, because the currents are small and change sign continuously. Diffusion does, however, contribute significantly at lower frequencies; solving

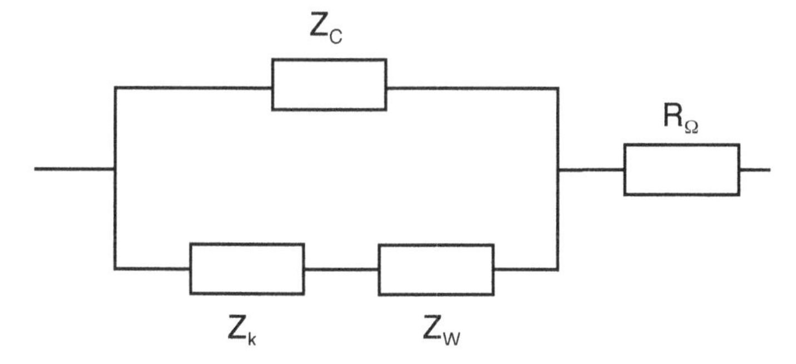

Fig. 19.7. Equivalent circuit for a simple redox reaction.

the diffusion equation with appropriate boundary conditions shows that the resulting impedance takes the form of the *Warburg impedance*:

$$Z_W = \frac{RT}{n^2F^2}\left(\frac{1}{c_{\rm red}D_{\rm red}^{1/2}} + \frac{1}{c_{\rm ox}D_{\rm ox}^{1/2}}\right)\frac{1-{\rm i}}{(2\omega)^{1/2}} \tag{19.24}$$

which is in series with Z_k, but parallel to Z_C. The resulting equivalent circuit is shown in Fig. 19.7, and in this simple case there is no ambiguity about the arrangement of the various elements.

There are several ways to plot the impedance spectrum $Z(\omega)$ or $Z(\nu)$. A common procedure is to plot the absolute value $|Z|$ of the impedance and the phase angle φ as a function of the frequency (see Fig. 19.8). In the example shown we chose values of: $R_\Omega = 1\ \Omega$, $C = 0.2$ F m^{-2}, $j_0 = 10^{-2}$ A cm^{-2}, diffusion coefficients of $D_{\rm ox} = D_{\rm red} = 5 \times 10^{-6}$ cm^2 s^{-1}, and concentrations of 10^{-2} M for both species. We assumed the presence of a supporting electrolyte with a higher concentration so that transport is by diffusion alone. At high frequencies the double-layer impedance Z_C is low and short circuits the charge-transfer branch. The impedance is then determined by the ohmic resistance R_Ω, and the phase angle is almost zero. At frequencies in the range of 10^3–10^4 Hz, most of the current flows through the capacitive branch. Therefore the phase angle is higher in this region. At lower frequencies Z_C is large, and the current flows mostly through the charge-transfer branch. The exchange current density can be evaluated from the data in the range of $10-10^3$ Hz. At lower frequencies transport is dominant, the current is determined by Z_W, and the phase angle rises towards 45°.

The form of such an impedance spectrum is readily understood if one realizes that it can be obtained from the current transient for a small potential step by Fourier transform. High frequencies correspond to short times, and low frequencies to long times. Thus double-layer charging dominates at short times and high frequencies, diffusion at long times and low frequencies.

For diagnostic purposes a plot of $-{\rm Im}(Z)$ versus ${\rm Re}(Z)$, a *Nyquist plot*, is useful, since certain processes give characteristic shapes. For example, the

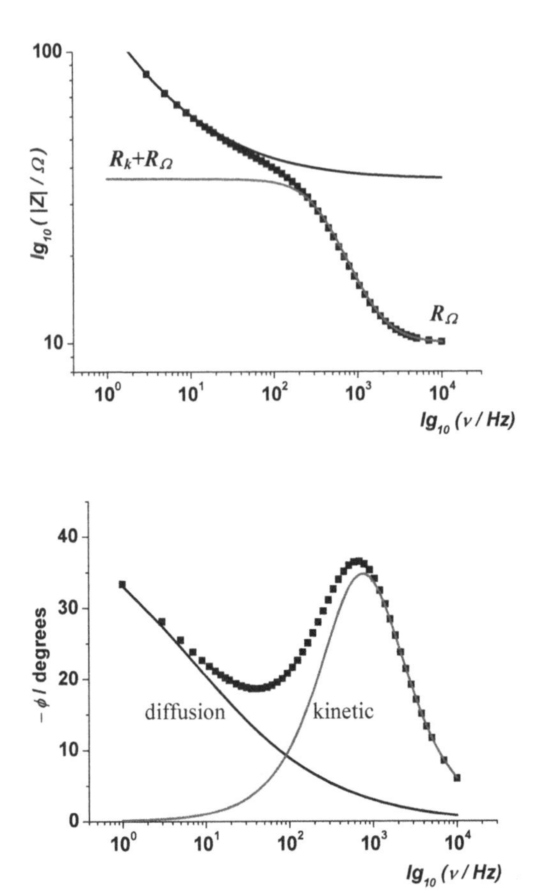

Fig. 19.8. Absolute value of the impedance and phase angle as a function of the frequency.

Warburg impedance shows up as a straight line with a slope of 45^0, a capacitor in parallel with a resistor gives a semicircle (see Problem 1). A simple charge-transfer reaction results in the beginning of a semicircle at high frequencies, which goes over into the Warburg line at low frequencies (see Fig. 19.9). When the charge transfer is fast, only a vestige of the semicircle can be seen.

Impedance spectroscopy is a good all-around method, giving both qualitative and quantitative information. It is easier to use than the pulse methods, but is limited to small deviations from equilibrium. Again, the upper limit of rate constants that can be measured is limited by double-layer charging, and is about the same as for the potential and current pulse methods.

19.7 Cyclic voltammetry

When faced with an unknown electrochemical system, or setting out on a new project, one generally starts with cyclic voltammetry. The electrode potential is varied cyclically and with a constant rate between two turning points (i.e., the applied potential varies in sawtooth-like fashion), and the current is recorded. Often the decomposition potentials of the solvent – for water, the onset potentials of hydrogen evolution and oxygen evolution – are chosen as turning points, but others may be chosen for special purposes. Sweep rates vary between a few mV s^{-1} up to $10^3 - 10^4$ V s^{-1}, depending on the purpose of the investigation. The resulting current-potential plot, the *cyclic voltammogram*, gives a survey over the processes occurring in the range studied.

As an example, Fig. 19.10 shows a cyclic voltammogram of a polycrystalline platinum electrode in 1 M H_2SO_4; it was recorded with a scan rate of 100 mV s^{-1}, a typical rate for the investigation of adsorption processes. Starting from 0 V vs. SHE, we see in the upper part of the curve, the posi-

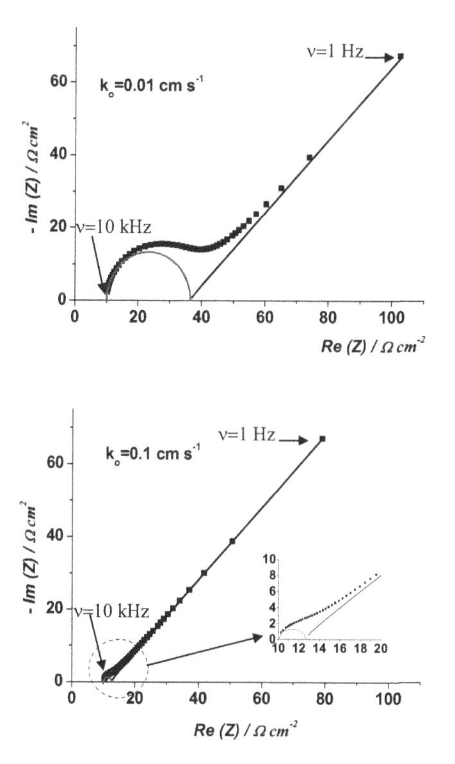

Fig. 19.9. Nyquist plot for a simple redox reaction for two different rate constants.

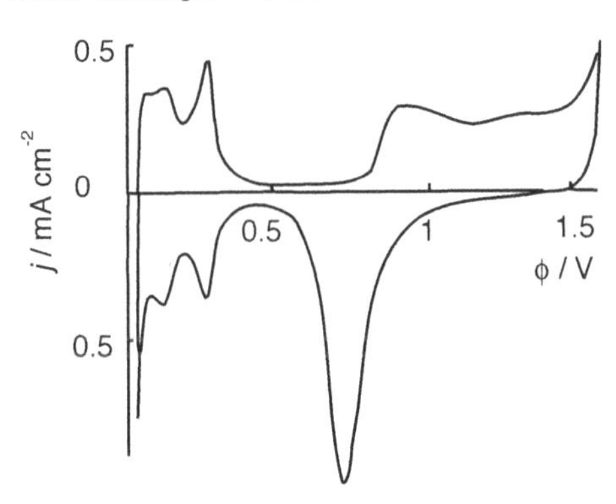

Fig. 19.10. Cyclic voltammogram of polycrystalline Pt in 1 M H_2SO_4 on SHE scale.

tive direction, first the desorption of adsorbed hydrogen; the different peaks correspond to different facets of single crystal surfaces on the polycrystalline material. At about 350 mV all hydrogen is desorbed, and the small residual current is due to double-layer charging. At about 850 mV PtO is formed at the surface, and oxygen evolution begins only at about 1.6 V, even though its thermodynamic equilibrium potential is at 1.23 V; as discussed in Sect. 13.3, its kinetics are slow and complicated. In the reverse sweep the PtO layer is desorbed; there is only a small double-layer region, and the adsorption of hydrogen begins again at 350 mV.

Polycrystalline metals are a badly defined superposition of various crystal faces. Actually, the response depends strongly on the surface structure and on the ions of the electrolyte. Figure 19.11 shows cyclic voltammograms of the three low index planes of Pt single crystals at a scan rate of 50 mV/s. The interpretation of the voltammogram of Pt(111) in sulfuric acid solution has been extensively discussed in the literature. The potential region of hydrogen underpotential adsorption, $0.07 < \phi < 0.3$ V, is clearly separated from the potential region for adsorption/desorption of bisulfate anions, $0.3 < \phi < 0.5$ V. At a more positive potentials, the OH_{ads} formation starts, which is hindered in the presence of adsorbing bisulfate anions. Cycling the electrode potential into the region where oxygen adsorption and desorption take place, leads to a successive disordering of the single crystal.

The characteristic features of the voltammogram of an ordered Pt(100) surface in sulfuric acid solution are two distinct peaks at 0.3 and 0.4 V, which mainly correspond to the coupled processes of hydrogen adsorption and bisulfate anion desorption on the (100) terrace sites and the (100) and (111) step sites, respectively. The potential region of H_{upd} is followed, first by the reversible adsorption of OH_{ads} in the potential range $0.7 < \phi < 0.85$ V, and then by the irreversible formation of platinum oxide at potentials more positive than 0.9 V.

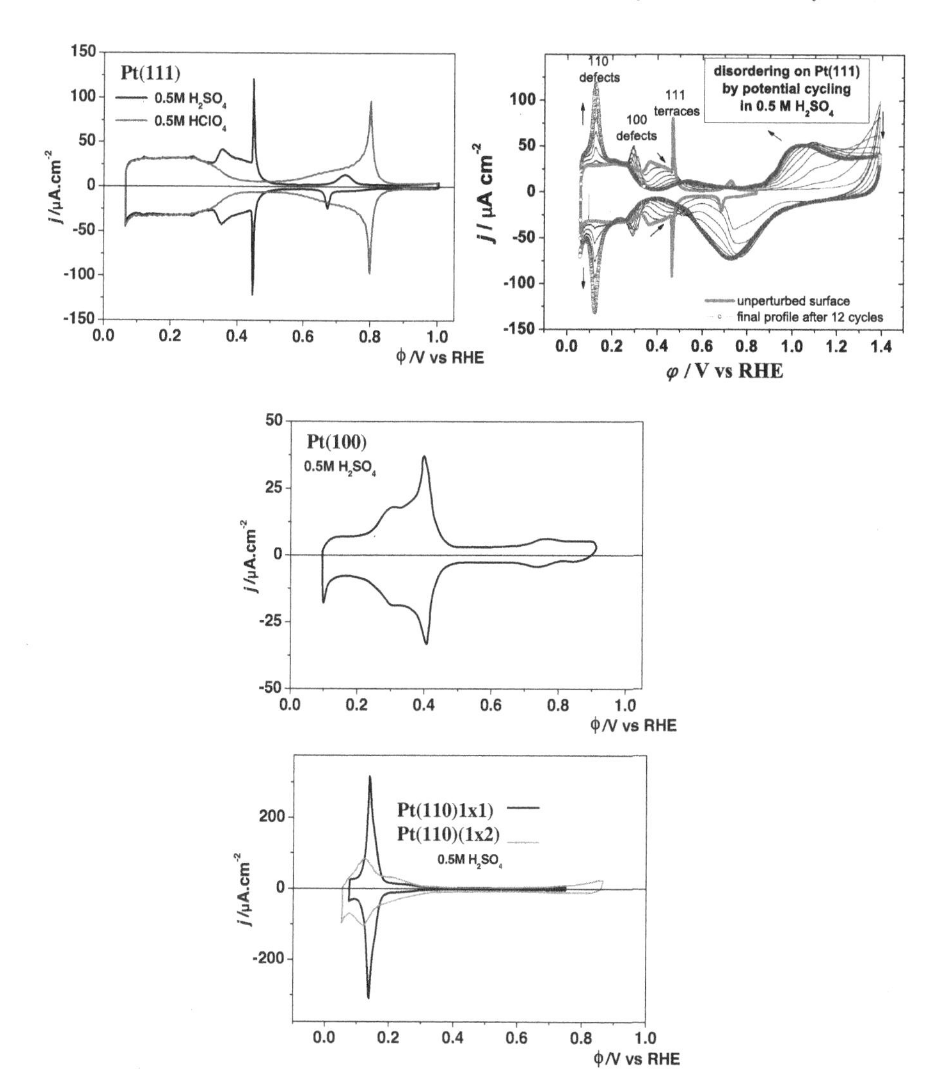

Fig. 19.11. Cyclic voltammogram of the three principle single-crystal surface of Pt on RHE scale. Data by courtesy of G. Beltramo, Jülich, and J. Feliu, Alicante [10]

In the case of Pt(110) surfaces, depending on the heat preparation treatment, it is possible to produce two different surface reconstructions. The 1×1 reconstruction can be produced by rapid gas-phase quenching (in argon with 3 % hydrogen), and the 1×2 or missing row reconstruction can be produced by slow cooling of the flame annealed crystal. The voltammograms of these two modifications differ significantly. The voltammetric features include reversible hydrogen adsorption/desorption peaks in the potential range of $0.05 - 0.35$ V, probably overlapping with bisulfate adsorption/desorption). Two peaks appear in the Pt(110)-(1×2) and are broader than the sharp peak observed in the Pt(110)-(1×1). These differences are attributed to the openness of the missing row structure.

Figure 19.12 shows voltammograms for gold single crystal electrodes. There is no detectable hydrogen adsorption region; the hydrogen evolution reaction is kinetically hindered, and sets in with a measurable rate only at potentials well below the thermodynamic value. There is a much wider double-layer region in which other reactions can be studied without interference. At higher positive potential we observe the formation of an oxide film, and its reduction in the negative sweep.

On both Au(111) and Au(100) the behavior is complicated by surface reconstruction, which has already been treated in Chap. 16. In particular the reconstruction of Au(100) entails a fairly large change in energy. In weakly adsorbing electrolytes it is lifted at potentials positive of the pzc, which is evidenced by a distinct peak in a slow cyclic voltammogram (see bottom panel). When the potential is scanned back towards negative potentials, the reconstruction is slow, and the corresponding peak is broader and not so high. Though the Au(111) surface is already densely packed, it exhibits a hexagonal reconstruction in the vacuum. Similarly to Au(100), this reconstruction is lifted at sufficiently positive potentials. Since the change in the surface structure is small, it only gives rise to small features in the voltammogram, which

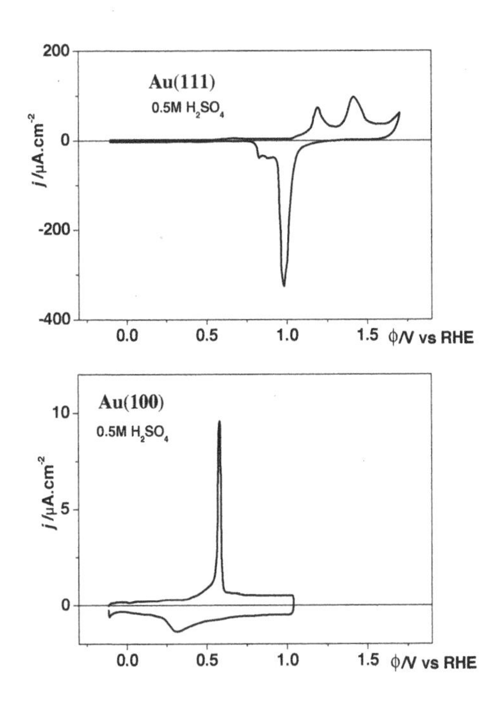

Fig. 19.12. Cyclic voltammograms of Au(111) and Au(100)

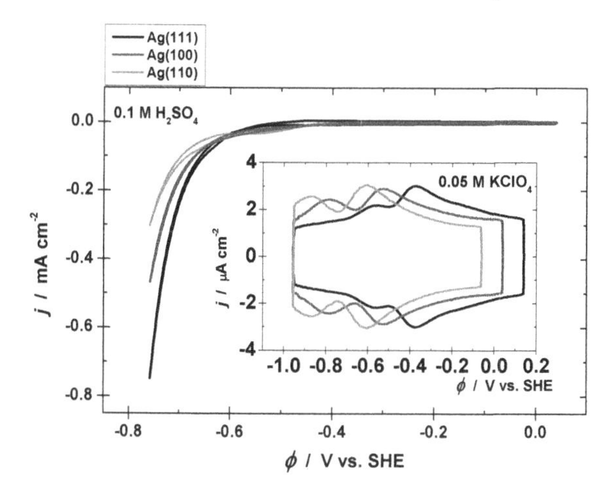

Fig. 19.13. Cyclic voltammograms of silver single-crystal surfaces. Data from [11].

fade between the peaks caused by the oxidation and reduction of the surface at potentials above 1 V.

Silver electrodes are interesting because of their wide double layer potential region and, contrary to gold, they do not show reconstruction processes. The cyclic voltammograms in Fig. 19.13 illustrate the electrochemical behavior of the different low index surface orientations. Similar to gold, the hydrogen evolution reaction is shifted to much more negative potentials than on platinum. There is a noticeable variation of the catalytic activity of the different surfaces. The inset of the figure shows the details of the double layer regions for a dilute, non-adsorbing electrolyte. The minima corresponding to the potential of zero charge can be easily distinguished.

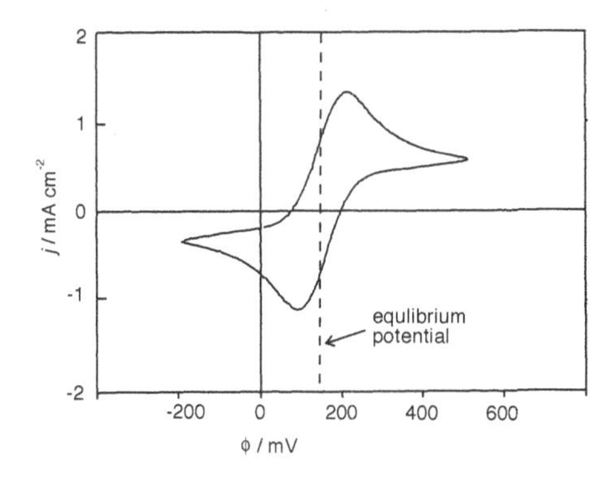

Fig. 19.14. Cyclic voltammogram of a simple electron transfer reaction.

A simple redox reaction shows a characteristic cyclic voltammogram exhibited in Fig. 19.14, which shows the situation after several cycles have already been performed, so that the original starting point has become irrelevant. In this example both the oxidized and the reduced species have the same concentrations in the bulk. We explain the shape of the curve for the positive sweep. At the lower left corner the potential is negative of the equilibrium potential, and a cathodic current is observed. Since this current has been flowing for some time, ever since the current became negative in this sweep, the concentration of the oxidized species at the surface is considerably lower than in the bulk. In the positive sweep the absolute magnitude of the overpotential, and hence also the cathodic current, become smaller, and the oxidized species is further depleted, while the reduced species is enriched. Therefore the current becomes zero at a potential below the equilibrium potential, and an anodic current starts to flow. With increasing potential, the rate of the anodic reaction becomes faster, and the current increases. However, simultaneously the reduced species is depleted at the surface, so that the current passes through a maximum, and becomes smaller as the surface concentration of the reduced species tends to zero. Usually the sweep direction is reversed soon after the maximum has been passed. Mutatis mutandis the same arguments can be used for the negative sweep.

This type of cyclic voltammogram is formed by the interplay of diffusion and the charge-transfer reaction; if the sweep rate is fast, double-layer charging also makes a significant contribution to the current. If the exchange current density and the transfer coefficient of the redox reaction, and furthermore the double-layer capacity, are known, the shape of the curve can be calculated numerically by solving the diffusion equation with appropriate boundary conditions. Conversely, these parameters can be determined from an experimental curve by a numerical fitting procedure. However, the curves are sensitive to the rate of the redox reaction only if the sweep rate is so fast that the reaction is not transport controlled throughout. For fast reactions this typically involves sweep rates of the order of 10^3 V s^{-1}. The whole procedure is useful only if the required computer programs are readily available. For slow reactions, as they often occur on organic electrochemistry, this is a suitable method, but not for fast reactions.

19.8 Microelectrodes

Spherical diffusion has peculiar properties, which can be utilized to measure fast reaction rates. The diffusion current density of a species i to a spherical electrode of radius r_0 is given by:

$$j_d = nFD_i c_i^0 \left(\frac{1}{(\pi D_i t)^{1/2}} + \frac{1}{r_0} \right) \tag{19.25}$$

The first term in the large parentheses is the same as that for a planar electrode, and it vanishes for $t \to \infty$. The second term is independent of time, so that a steady diffusion current is obtained after an initial period. Even though the region near the electrode gets more and more depleted as the reaction proceeds, material is drawn in from an ever-increasing region of space, and these two effects combine to give a constant gradient at the electrode surface. By making the radius of the electrode sufficiently small, the diffusion current density can be made arbitrarily large, as large as the kinetic current of any electrochemical reaction, so that any rate constant could, in principle, be measured!

There are, however, obvious limitations. It is not possible to make a very small spherical electrode, because the leads that connect it to the circuit must be even much smaller lest they disturb the spherical geometry. Small disc or ring electrodes are more practicable, and have similar properties, but the mathematics becomes involved. Still, numerical and approximate explicit solutions for the current due to an electrochemical reaction at such electrodes have been obtained, and can be used for the evaluation of experimental data. In practice, ring electrodes with a radius of a fraction of a μm can be fabricated, and rate constants of the order of a few cm s^{-1} be measured by recording currents in the steady state. The rate constants are obtained numerically by comparing the actual current with the diffusion-limited current.

Even though their fabrication is difficult, microelectrodes have a number of advantages over other methods:

1. Since measurements can be performed in the steady state, double-layer charging plays no role.
2. Only small amounts of solutions and reactants are required.
3. Currents are small, and so is the IR drop between the working and the reference electrode, so that microelectrodes are particularly useful in solutions with a low conductivity.
4. Because of their small size, they can be used in biological systems.

19.9 Complementary methods

The methods described above rely on the measurements of current and potential, and provide no direct information about the microscopic structure of the interface, though a clever experimentalist may make some inferences. During the past 30 years a number of new techniques have been developed that allow a direct study of the interface. This has led to substantial progress in our understanding of electrochemical systems, and much more is expected in the future. Thus we have the possibility of applying additional perturbations to the interface, which provide complementary information to the classical electrochemical variables such as potential, current and charge.

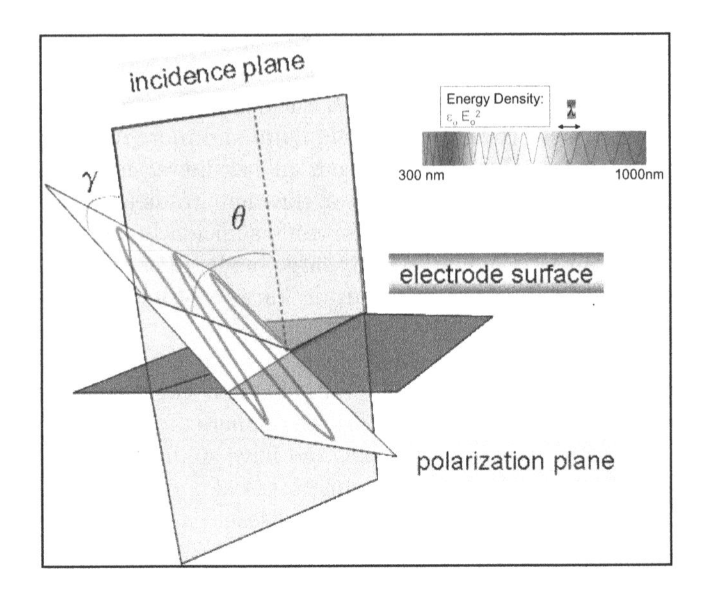

Fig. 19.15. Interaction of light with an electrode surface.

A particular interesting perturbation is light, which implies the presence of oscillating electromagnetic fields at the interface. Thus, besides the double-layer field we have an additional electrical field, whose direction we can change in a simple way by changing the polarisation angle γ of the light as shown in Fig. 19.15. Changing the intensity of the light, we can investigate both linear and non linear phenomena. By changing the wavelength of the light we change the frequency of the oscillating fields, but in contrast to impedance spectroscopy the range is now within $10^{14} - 10^{15}$ Hz. So, we can follow much faster processes with time constants of the order of 1–10 fs. Resonance phenomena corresponding to processes such as electronic and vibronic transitions can be easily identified. Many of these methods are variants of spectroscopies familiar from other fields.

All methods in which the electrode surface is investigated as it is, in contact with the solution, are called in situ methods. In ex situ methods the electrode is pulled out of the solution, transferred to a vacuum chamber, and studied with surface science techniques, in the hope that the structure under investigation, such as an adsorbate layer, has remained intact. Ex situ methods should only be trusted if there is independent evidence that the transfer into the vacuum has not changed the electrode surface. They belong to the realm of surface science, and will not be considered here.

Problems

1. Consider the impedance circuit of Fig. 19.7. Show that for $Z_W = 0$ a Nyquist plot gives a semicircle. If $Z_W \neq 0$ calculate the frequency region in which the semicircle merges into a straight line of unit slope.
2. From Eq. (19.4) derive an asymptotic expression for the current density which is valid in the region $\lambda t^{1/2} \gg 1$.
3. Consider the generation of a species at a spherical electrode. In polar coordinates the diffusion equation is:

$$\frac{\partial c}{\partial t} = \frac{D}{r^2}\frac{\partial}{\partial r}\left(\frac{r^2 \partial c}{\partial r}\right) \tag{19.26}$$

Show that this equation has a steady-state solution, and derive a general expression for the concentration and the diffusion current.

References

1. Southampton Electrochemistry Group, *Instrumental Methods in Electrochemistry*, Ellis Horwood Limited, Chichester, 1985.
2. A.J. Bard and L.R. Faulkner, *Electrochemical Methods: Fundamentals and Applications*, 2nd edition. Wiley, New York, NY, 2001.
3. D.D. MacDonald, *Transient Techniques in Electrochemistry*. Plenum Press, New York, NY, 1977.
4. P. Delahay, *New Instrumental Methods in Electrochemistry*. Wiley-Interscience, New York, NY, 1954.
5. E. Gileadi, *Electrode Kinetics*. VCH, New York, NY, 1993.
6. Z. Galus, *Fundamentals of Electrochemical Analysis*, 2nd edition. Ellis Horwood, New York, NY, 1994.
7. P. Delahay, *J. Phys. Chem.* **66** (1962) 2204.
8. *Analyt. Chem.* **34** (1962) 1272.
9. E. Santos, T. iwasita, and W. Vielstich, *Electrochim. Acta* **31** (1986) 431.
10. A. Björling, E. Ahlberg, and J.M. Feliu, *Electrochem. Commun.* **12** (2010) 359.
11. D. Eberhardt, Ph.D. Thesis, Ulm University, 2001.

20

Convection techniques

Forced convection can be used to achieve fast transport of reacting species toward and away from the electrode. If the geometry of the system is sufficiently simple, the rate of transport, and hence the surface concentrations c^s of reacting species, can be calculated. Typically one works under steady-state conditions so that there is no need to record current or potential transients; it suffices to apply a constant potential and measure a stationary current. If the reaction is simple, the rate constant and its dependence on the potential can be calculated directly from the experimental data.

Working under steady-state conditions has certain advantages; in particular the complications caused by double-layer charging are avoided. On the other hand, convection techniques require a greater volume of solution, and contamination of the electrode surface is even more of a problem than usual because the solution is constantly swept past the electrode surface.

20.1 Rotating disc electrode

The simplest and most commonly used convection apparatus consists of a disc electrode rotating with a constant angular velocity ω [1–5]. The disc sucks the solution toward its surface, much in the way a propeller would; as the solution approaches the disc, it is swept away radially and tangentially (see Fig. 20.1). The transport of the reacting species to the disc occurs both by convection and diffusion. Though the mathematics are complicated, the rate of transport can be calculated exactly for an infinite disc. A particularly nice feature of this setup is the fact that the transport is uniform so that the surface concentration of any reacting species is constant over the surface of the electrode.

Right at the disc the convection current perpendicular to the surface vanishes. The transport to the surface is effected by diffusion; so the particle current density j_p of any species with concentration c and diffusion coefficient D toward the electrode is:

W. Schmickler, E. Santos, *Interfacial Electrochemistry*, 2nd ed.,
DOI 10.1007/978-3-642-04937-8_20, © Springer-Verlag Berlin Heidelberg 2010

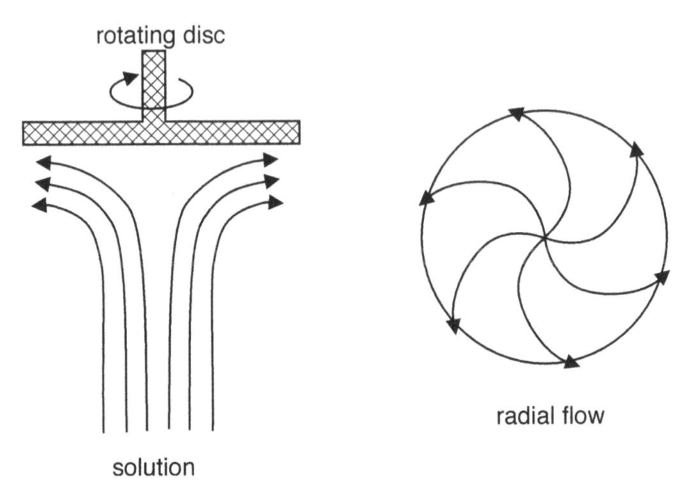

Fig. 20.1. Convection current at a rotating disc electrode.

$$j_p = -D \left(\frac{dc}{dx} \right)_{x=0} \tag{20.1}$$

As mentioned above, on the disc this current is independent of position. It is useful to define a *diffusion layer* of thickness δ_N through:

$$\left(\frac{dc}{dx} \right)_{x=0} = \frac{c^b - c^s}{\delta_N} \tag{20.2}$$

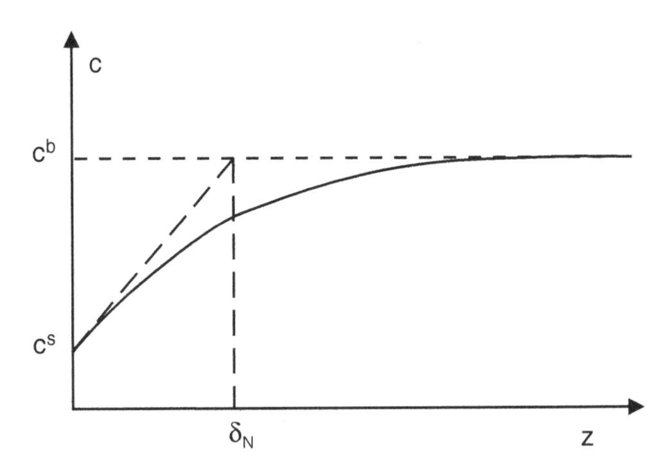

Fig. 20.2. Definition of the diffusion layer thickness δ_N.

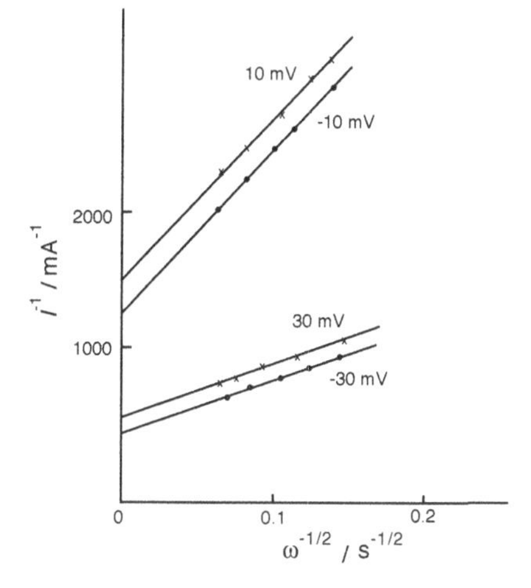

Fig. 20.3. Koutecky–Levich plot for four different overpotentials.

where c^b and c^s are the bulk and the surface concentrations of the diffusing species. For many purposes one may replace the complicated concentration profile of a diffusing species by a simplified one, in which the concentration gradient is constant within the diffusion layer, and the concentration itself is constant and equal to the bulk concentration in the region beyond (see Fig. 20.2.

At a rotating disc the thickness of the diffusion layer decreases with increasing rotation rate according to [1, 2]:

$$\delta_N = 1.61 D^{1/3} \nu^{1/6} \omega^{-1/2} \left[1 + 0.35 \left(\frac{D}{\nu} \right)^{0.36} \right] \tag{20.3}$$

where ν is the *kinematic viscosity* of the solution, which is obtained from the usual dynamic viscosity ζ by $\nu = \zeta/\rho$, where ρ is the density; the numerical constants follow from the complicated mathematics of the equations for convective diffusion. For a simple redox reaction the current density is (see Chap. 9):

$$j = F \left(k_{\text{ox}} c_{\text{red}}^s - k_{\text{red}} c_{\text{ox}}^s \right) \tag{20.4}$$

Under steady-state conditions each molecule "red" transported to the surface is oxidized, and hence transformed to "ox"; hence $j/F = j_p^{\text{red}} = -j_p^{\text{ox}}$, or:

$$j = F D_{\text{red}} \frac{c_{\text{red}}^b - c_{\text{red}}^s}{\delta_{\text{red}}} = -F D_{\text{ox}} \frac{c_{\text{ox}}^b - c_{\text{ox}}^s}{\delta_{\text{ox}}} \tag{20.5}$$

using an obvious notation. Equations (20.4) and (20.5) can be recast in the form:

$$\frac{1}{j} = \frac{1}{j_k} + B \omega^{-1/2} \tag{20.6}$$

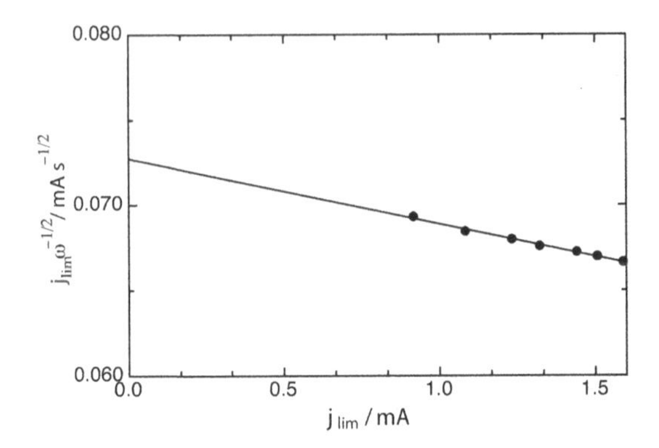

Fig. 20.4. Plot for the evaluation of the dissociation constant.

where B is a constant depending on the diffusion coefficients, the viscosity, and the reaction rate; j_k is the kinetic current density, which would flow if the concentration at the surface were the same as in the bulk. A plot of $1/j$ versus $\omega^{-1/2}$, a so-called *Koutecky–Levich plot*, gives a straight line with intercept $1/j_k$ (see Fig. 20.3). This is an extrapolation to infinitely fast mass transport, for which surface and bulk concentrations would be equal, and the measured current j would equal the kinetic current j_k.

The current-potential characteristics of a redox reaction can thus be measured in the following way: An overpotential η is applied, and the current is measured for various rotation rates ω. From a Koutecky–Levich plot the corresponding kinetic current $j_k(\eta)$ is extrapolated. This procedure is repeated for a series of overpotentials, and the dependence of j_k on η is determined.

There are several variants of this method for more complicated reactions. If the reacting species is produced by a preceding chemical reaction, deviations from Eq. (20.6) may be observed for large ω, when the reaction is slower than mass transport. From these deviations the rate constant of the chemical reaction can be determined. As an example we consider hydrogen evolution from a weak acid HA, where the reacting protons are formed by a preceding dissociation reaction:

$$\begin{aligned} \mathrm{HA} &\rightleftharpoons \mathrm{H^+ + A^-} \\ \mathrm{2H^+ + 2e^-} &\rightarrow \mathrm{H_2} \end{aligned} \tag{20.7}$$

Equation (20.6) predicts that for constant ω and large overpotential η the current becomes equal to $B\omega^{-1/2}$; under these circumstances j_k is very large, the current is transport controlled, the surface concentration is negligible, and $j = j_{\mathrm{lim}} = B\omega^{-1/2}$, the limiting current density. This remains true for the scheme of Eq. (20.7) as long as the dissociation reaction is faster than mass transport. However, for large ω dissociation can no longer supply the protons at the required rate, and the limiting current is determined by dissociation

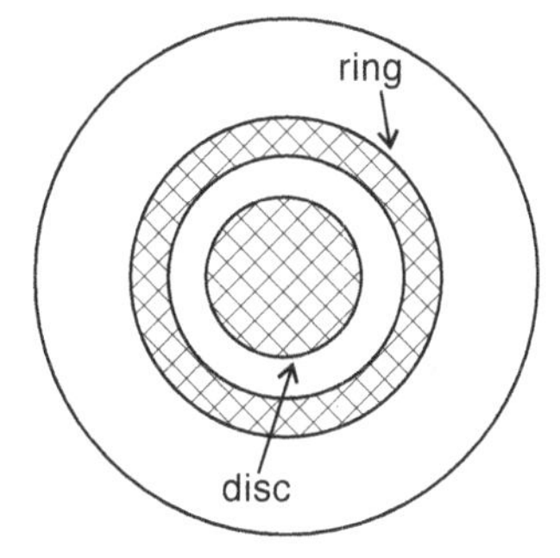

Fig. 20.5. Rotating ring-disc electrode.

and not by mass transport. If the concentration c_{A^-} of the anions A^- in the bulk is much larger than that of the protons, an explicit formula can be derived [3–5]:

$$j_{\lim}\omega^{-1/2} = j_{tr}\omega^{-1/2} - \frac{D^{1/6} c_{A^-} j_{\lim}}{1.61\nu^{1/6}(k_d/k_r)k_r^{1/2}} \tag{20.8}$$

j_{tr} is the limiting current for an infinitely fast dissociation reaction, k_d the rate constant of the dissociation, and k_r that of the recombination. A plot of $j_{\lim}\omega^{-1/2}$ versus $j_{\lim}$ gives a straight line (see Fig. 20.4), and the rate constant k_r of the dissociation reaction can be determined from the slope, if the diffusion coefficient D of the protons, the kinematic viscosity ν, and the dissociation constant k_d/k_r are known or determined by separate measurements. A well-known example is the dissociation of acetic acid with a rate constant of about 5×10^5 s^{-1}; so this is one of the faster methods to measure rate constants of preceding chemical reactions.

Another extension of this method is the use of a concentric ring surrounding the disc, at which intermediate products can be determined, a so-called *ring-disc electrode* (see Fig. 20.5). Any species that is generated at the central disc is swept past the ring. Due to the particular hydrodynamics of this system, the *collection efficiency* N, which is defined as the fraction of a stable species generated at the disc that reaches the ring, depends on the geometry of the electrodes only and is independent of the rotation rate. Typically, N is of the order of 20% or more.

While N can be calculated for a given geometry, it is usually determined experimentally by using a simple electron-transfer reaction. A species is oxidized at the disc and reduced at the ring (or vice versa):

$$\begin{aligned} \mathrm{A} &\rightleftharpoons \mathrm{B} + \mathrm{e}^- \quad \text{(disc)} \\ \mathrm{B} + \mathrm{e}^- &\rightleftharpoons \mathrm{A} \qquad \text{(ring)} \end{aligned} \tag{20.9}$$

If the potential at the ring is chosen such that the ring current is transport limited, then:

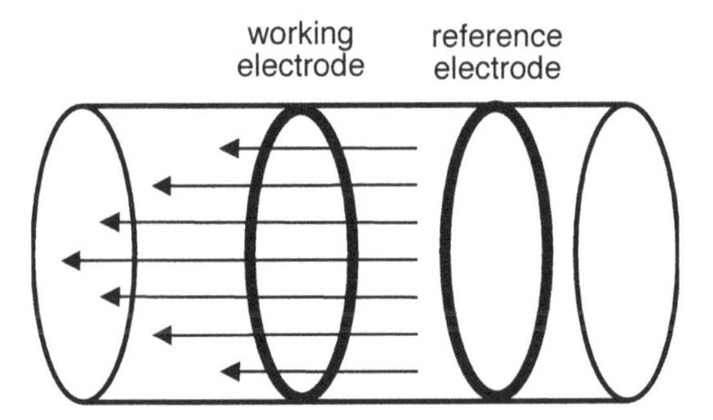

Fig. 20.6. Section of a pipe for turbulent flow.

$$I_{\text{ring}} = N I_{\text{disc}} \tag{20.10}$$

provided the oxidized species does not react while it is transported from the disc to the ring. Ring-disc electrodes have been used successfully in the study of reactions such as the oxygen reduction. We consider a particular example in the problems. For further details we refer to [3, 4].

20.2 Turbulent pipe flow

The faster the flow of the solution, the faster the mass transport, and the higher the reaction rates that can be measured. The *Reynolds number Re*, defined as $Re = vL/\nu$, where v is the velocity of flow and L a characteristic length such as the diameter of a pipe, is a convenient dimensionless quantity to characterize the rate of flow in various systems. In a cylindrical pipe the flow is turbulent for Reynolds numbers $Re > 2,000$, and mass transport is particularly fast. If the working electrode is cast in the form of a ring embedded in the wall of the pipe (see Fig. 20.6), mass transport is fastest at the front edge facing the flow, because the reactants are depleted downstream. So thin rings are particularly suitable for kinetic investigations. On the other hand, the ring never fits quite smoothly into the wall of the pipe, and the resulting edge effects will distort the flow seriously if the ring is too thin. In practice, a ring thickness of the order of 50 – 100 μm is a good compromise.

In contrast to the rotating disc electrode, mass transport to the ring is nonuniform. Nevertheless, the thickness of the diffusion layer δ_N, which depends on the coordinate x in the direction of flow, and the rate of mass transport can be calculated. We consider a simple redox reaction, and rewrite Eq. (20.5) in the form:

$$\begin{aligned} j &= F D_{\text{red}} \frac{c^b_{\text{red}} - c^s_{\text{red}}}{\delta_{\text{red}}} = a_{\text{red}} \left(c^b_{\text{red}} - c^s_{\text{red}} \right) \\ &= -F D_{\text{ox}} \frac{c^b_{\text{ox}} - c^s_{\text{ox}}}{\delta_{\text{ox}}} = -a_{\text{ox}} \left(c^b_{\text{ox}} - c^s_{\text{ox}} \right) \end{aligned} \tag{20.11}$$

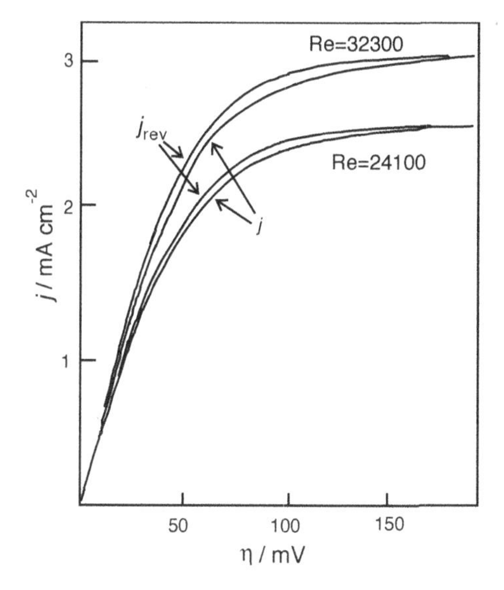

Fig. 20.7. Current-potential curves for two different Reynolds numbers.

where we have introduced an obvious notation. We first consider the reversible current. When the overpotential is very high, the concentration of the reduced species at the surface vanishes, and the corresponding anodic limiting current density is:

$$j_{\text{lim}}^a = a_{\text{red}} c_{\text{red}}^b \tag{20.12}$$

Similarly the cathodic limiting current is:

$$j_{\text{lim}}^c = -a_{\text{ox}} c_{\text{ox}}^b \tag{20.13}$$

If the electron-transfer reaction were infinitely fast, the overpotential would be given by Nernst's equation in the form:

$$\eta = \frac{RT}{F} \left(\ln \frac{c_{\text{ox}}^s}{c_{\text{red}}^s} - \ln \frac{c_{\text{ox}}^b}{c_{\text{red}}^b} \right) \tag{20.14}$$

Substituting Eqs. (20.12),(20.13), and(20.14) into Eq. (20.11) gives for the reversible current density:

$$j_{\text{rev}} = \frac{j_{\text{lim}}^c j_{\text{lim}}^a \left[1 - \exp\left(-F\eta/RT\right)\right]}{j_{\text{lim}}^c - j_{\text{lim}}^a \exp\left(-F\eta/RT\right)} \tag{20.15}$$

We note in passing that the same equation holds for the rotating disc electrode. Though the mass transport on the ring is nonuniform, the ratio $a_{\text{red}}/a_{\text{ox}}$, and hence also $j_{\text{lim}}^a/j_{\text{lim}}^c$, turns out to be constant, so Eq. (20.15) remains valid if we substitute the currents $I_{\text{rev}}, I_{\text{lim}}^a, I_{\text{lim}}^c$ for the current densities. Solving the mass transport explicitly – a nontrivial task for turbulent flow – shows that the measured current is given by [6]:

$$I = I_{\text{rev}} \left[1 - 2u + 2u^2 \ln(1 + 1/u)\right] \tag{20.16}$$

where the dimensionless parameter $u = 2I_{\text{rev}}/I_k$ contains the kinetic current.

The experiment is evaluated in the following way: For a given flow rate the current I is measured as a function of the applied overpotential η (see Fig. 20.7), and the limiting currents at high anodic and cathodic overpotentials are obtained. Then the reversible current I_{rev} is calculated from Eq. (20.15). From I_{rev} and from the measured I the parameter u, and hence the kinetic current I_k, is obtained by solving Eq. (20.16) numerically. The faster the mass transport, the larger the difference between I and I_{rev}, and the more precise is the measurement of I_k. This technique has the same advantages and disadvantages as the rotating disc, but mass transport is appreciably faster, and rate constants up to 5 cm s^{-1} can be measured.

Problems

1. From Eqs.(20.1), (20.2), and (20.3), show that the particle current density j_p of a species A at a rotating disc can be written in the form:

$$j_p^A = \gamma_A \omega^{1/2} \left(c_A^b - c_A^s\right) \tag{20.17}$$

where γ_A is a constant that depends on the diffusion coefficient of the species A and on the kinematic viscosity of the solution. Consider a ring-disc electrode at which a reaction of the form of Eq. (20.9) takes place. Show that the ring current is:

$$I_{\text{ring}} = NSFj_p^B = NSF\gamma_B \omega^{1/2} c_B^s \tag{20.18}$$

where S is the area of the disc, and j_p^B and c_B^s are the particle current density and the concentration at the disc.

2. We consider the investigation of two consecutive electron-transfer reactions with a ring-disc electrode under stationary conditions. A species A reacts in two steps on the disk electrode: first to an intermediate B which reacts further to the product C. The intermediate is transported to the ring, where the potential has been chosen such that it reacts back to A. The overall scheme is:

disc: $A \xrightarrow[n_1]{k_1} B \xrightarrow[n_2]{k_2} C$ (with B transported ↓ to the ring)

ring: $B \xrightarrow[n_1]{} A$

The rate constants and the number of electrons transferred are indicated in the diagram. The back reaction at the ring is supposed to be so fast that every molecule of B that reaches the ring is immediately consumed. Further,

B is supposed to be absent from the bulk of the solution. The current at the disc is:

$$I_{\text{disc}} = n_1 F S k_1 c_A^s + n_2 F S k_2 c_B^s \tag{20.19}$$

where S is the area of the disc. Write down the mass balance conditions for the species A and B at the disc. Show that:

$$N \frac{I_{\text{disc}}}{I_{\text{ring}}} = 1 + \frac{n_1 + n_2}{n_1} \frac{k_2}{\gamma_B \omega^{1/2}} \tag{20.20}$$

The limiting current at the disc is:

$$I_{\text{lim}} = (n_1 + n_2) S F \gamma_A c_A^b \omega^{1/2} \tag{20.21}$$

Show that:

$$n_1 N \frac{I_{\text{lim}} - I_{\text{disc}}}{I_{\text{ring}}} = n_2 + (n_1 + n_2) \frac{\gamma_A k_2}{\gamma_B k_1} + (n_1 + n_2) \frac{\gamma_A \omega^{1/2}}{k_1} \tag{20.22}$$

How can the rate constants k_1 and k_2 be obtained by a suitable experiment?

References

1. V.G. Levich, *Physicochemical Hydrodynamics.* Prentice Hall, Englewood Cliffs, NJ, 1962.
2. A.C. Riddiford, *Advances in Electrochemistry and Electrochemical Engineering,* Vol. 4, edited by P. Delahay and C.W. Tobias. Wiley, New York, NY, 1966.
3. W.J. Albery and M.L. Hitchman, *Ring-Disc Electrodes.* Clarendon Press, Oxford, 1971.
4. Yu.V. Pleskov and V.Yu. Filinowskii, *The Rotating Disc Electrode.* Consultants Bureau, New York, NY, 1976.
5. J.S. Newman, *Electrochemical Systems.* Prentice Hall, Englewood Cliffs, NJ, 1962.
6. F. Barz, C. Bernstein, and W. Vielstich, *Advances in Electrochemistry and Electrochemical Engineering,* Vol. 13, edited by H. Gerischer and C.W. Tobias. Wiley, New York, NY, 1984.

Index

W. Schmickler, E. Santos, *Interfacial Electrochemistry*, 2nd ed.,
DOI 10.1007/978-3-642-04937-8, © Springer-Verlag Berlin Heidelberg 2010